Administrator's Guide

How to Support and Improve
Mathematics Education in Your School

Amy J. Mirra

NATIONAL COUNCIL OF
TEACHERS OF MATHEMATICS

RESTON, VIRGINIA

Association for Supervision
and Curriculum Development

Alexandria, Virginia

Copyright © 2003 by
THE NATIONAL COUNCIL OF TEACHERS OF MATHEMATICS, INC.
1906 Association Drive, Reston, VA 20191-1502
(703) 620-9840; (800) 235-7566; www.nctm.org

Fourth printing 2006

ISBN 0-87353-552-9

The National Council of Teachers of Mathematics is a public voice of mathematics education,
providing vision, leadership, and professional development to support teachers in ensuring
mathematics learning of the highest quality for all students.

Printed in the United States of America

Table of Contents

Preface

OVER THE PAST DECADE, the National Council of Teachers of Mathematics (NCTM) has advocated for changes in school mathematics programs so that all students have the opportunity to engage in high-quality mathematics that will prepare them for today and a world tomorrow they can only imagine. NCTM's *Principles and Standards for School Mathematics* (2000) describes six principles and ten standards to guide mathematics instruction from prekindergarten through grade 12. Realizing the vision for mathematics education presented in *Principles and Standards* is a formidable task, but it is an educational imperative for the future welfare of our students. It requires solid "mathematics curricula, competent and knowledgeable teachers who can integrate instruction with assessment, education policies that enhance and support learning, classrooms with ready access to technology, and a commitment to both equity and excellence" (NCTM 2000, p. 3). It also requires the active participation not only of teachers but of administrators, policymakers, higher-education faculty, curriculum developers, researchers, families, students, and community members.

This guide presents many of the ideas from *Principles and Standards* and previous NCTM *Standards* documents. It also includes some of the research cited in *Principles and Standards*, as well as other research and reports on mathematics

teaching and learning. Although some of the content is written specifically for a building-level administrator, much of it is appropriate for all administrators. Administrators at every level play a role in the improvement of mathematics education and should be aware of the fundamental issues and ways to support these changes.

The guide begins by discussing what it means to be mathematically literate and presents examples from an elementary, middle, and high school classroom to give the reader a picture of what might constitute a high-quality classroom. The six NCTM principles—Equity, Curriculum, Teaching, Learning, Assessment, and Technology—are then presented as the basic precepts that are fundamental for a high-quality mathematics program. These principles offer perspectives that can guide decision making in mathematics education. From there, the guide moves on to specific actions that administrators can take to support mathematics education in their school or schools. The section "Frequently Asked Questions" offers guidance on questions that are often asked in mathematics education. Finally, the guide presents a list of resources and other sources of assistance. In time, with this guidance, you can begin to make a difference in the quality of teaching and learning of mathematics in your school.

Acknowledgments

The National Council of Teachers of Mathematics wishes to acknowledge the extensive efforts of Amy Mirra, Communications Outreach Coordinator, in gathering and writing the information for this guide. The Council also extends a special thank-you to the following people, who reviewed and provided feedback on the content of this document:

External Reviewers

Suzanne Austin, Miami-Dade Community College, Miami, Florida

Michael Haynes, East Jordan Middle School, East Jordan, Michigan

Virginia Horak, University of Arizona, Tucson, Arizona

Rita Janes, Mathematics Consultant, Saint John's, Newfoundland

Donna Jenner, University of British Columbia, Vancouver, British Columbia

Jeane Joyner, Meredith College, Raleigh, North Carolina

Valerie Mills, Ann Arbor School District, Ann Arbor, Michigan

Vodene Schultz, El Paso Independent School District, El Paso, Texas

George Viebranz, Berea City School District, Berea, Ohio

Barb Younkin, Berea City School District, Berea, Ohio

NCTM Staff

Marilyn Hala

Ken Krehbiel

Lynda McMurray

Rebecca Ries

Beth Skipper

Helen Snyder

Harry Tunis

Fundamental Ideas for High-Quality Mathematics Education

- Mathematical literacy is essential for every child's future.

- A solid mathematics education is essential for an informed public, our national security, a strong economy, and national well-being.

- All students can be successful in mathematics and should receive a high-quality mathematics education, regardless of gender, ethnicity, or race.

- Teachers should encourage and inspire every student to continue the study of mathematics.

- Developing mathematical proficiency requires a balance and connection between conceptual understanding and procedural and computational proficiency.

- Problem solving and using mathematics to understand our world is an integral part of all mathematics learning.

- Teachers must have a solid knowledge of both mathematics content and teaching strategies as well as enjoy and value mathematics.

- Effective programs of teacher preparation and professional development help teachers understand the mathematics they teach, how their students learn that mathematics, and how to help each student learn.

- Improving mathematics education for all requires a commitment from a variety of stakeholders, including teachers, mathematics teacher-leaders, school and district administrators, institutions of higher learning, mathematicians, professional organizations, families, politicians, business and community leaders, and students.

Introduction

The world today is much different from that of even a few years ago. We are all bombarded with data that must be absorbed, sorted, organized, and used to make increasingly crucial decisions. The underpinnings of everyday life, such as making purchasing decisions, choosing insurance or health plans, and planning for retirement, all require mathematical competence. Business and industry demand workers who can solve real-world problems, explain their thinking to others, identify and analyze trends in data, and use modern technology.

Recent data from the Bureau of Labor and Statistics also reveal that more students must pursue paths in mathematical and technical occupations. Employment projections to 2010 expect these occupations to add the most jobs, 2 million, and grow the fastest among the eight professional and related occupational subgroups (Hecker 2001). But will enough talent be available to fill these positions? Another startling statistic says that 60 percent of all new jobs in the early twenty-first century will require skills that are possessed by only 20 percent of the current workforce (National Commission on Mathematics and Science for the Twenty-first Century 2000).

> **Business and industry demand workers who can solve real-world problems, explain their thinking to others, identify and analyze trends in data, and use modern technology.**

Snapshot of How Students Are Performing in Mathematics

National and international studies of students' achievement in mathematics in the United States are both discouraging and hopeful. Overall, students in the United States are not performing well enough in mathematics. However, these tests also show that steady improvement has occurred over the last decade.

The results of the 2000 National Assessment of Educational Progress (NAEP), which ranks students' performance using four levels of achievement—below basic, basic, proficient, and advanced—showed the following major results (National Center for Education Statistics [NCES] 2003a):

- Students in all three grades (4, 8, and 12) had higher average scores in 2000 than in 1990.

- The percent of fourth graders at or above the proficient level doubled, from 13 percent in 1990 to 26 percent in 2000.

- The percent of fourth graders at or above the basic level increased from 50 percent in 1990 to 69 percent in 2000.

- The percent of eighth graders at or above the proficient level increased from 15 percent in 1990 to 27 percent in 2000.

- The percent of eighth graders at or above the basic level increased from 52 percent in 1990 to 66 percent in 2000.

- The percent of twelfth graders at or above the proficient level increased from 12 percent in 1990 to 17 percent in 2000.

- The percent of twelfth graders at or above the basic level increased from 58 percent in 1990 to 65 percent in 2000.

The Third International Mathematics and Science Study (TIMSS) in 1995 and the Third International Mathematics and Science Study Repeat (TIMSS-R) in 1999 echo many of the same findings from NAEP. Overall, the TIMSS has shown that U.S. students have maintained their standing in the middle of the international ranking. Canada experienced a significant increase in its students' mathematics average between the 1995 and 1999 TIMSS. In fact, Canada was one of only three countries to show significant score gains between the two studies. The United States and Canada, however, are both significantly below the top-performing countries (NCES 2003b, 2003c).

PURPOSE OF THIS GUIDE

The importance of mathematical literacy and the need to understand and be able to use mathematics in everyday life and in the workplace have never been greater and will continue to increase. Meeting this demand for all students implies changes in curricular expectations for students as well as in instructional practices. This booklet provides a starting place for what you, as an administrator, need to know about mathematics education and ways you can support and improve mathematics education in your school.

Average mathematics and science achievement of eighth-grade students, by nation: 1999

MATHEMATICS		SCIENCE	
Nation	**Average**	**Nation**	**Average**
Singapore	604	Chinese Taipei	569
Korea, Republic of	587	Singapore	568
Chinese Taipei	585	Hungary	552
Hong Kong SAR	582	Japan	550
Japan	579	Korea, Republic of	549
Belgium-Flemish	558	Netherlands	545
Netherlands	540	Australia	540
Slovak Republic	534	Czech Republic	539
Hungary	532	England	538
Canada	531	Finland	535
Slovenia	530	Slovack Republic	535
Russian Federation	526	Belgium-Flemish	535
Australia	525	Slovenia	533
Finland[1]	520	Canada	533
Czech Republic	520	Hong Kong SAR	530
Malaysia	519	Russian Federation	529
Bulgaria	511	Bulgaria	518
Latvia-LSS[2]	505	United States	515
United States	502	New Zealand	510
England	496	Latvia-LSS[2]	503
New Zealand	491	Italy	493
Lithuania[3]	482	Malaysia	492
Italy	479	Lithuania[3]	488
Cyprus	476	Thailand	482
Romania	472	Romania	472
Moldova	469	(Israel)	468
Thailand	467	Cyprus	460
(Israel)	466	Moldova	459
Tunisia	448	Macedonia, Republic of	458
Macedonia, Republic of	447	Jordan	450
Turkey	429	Iran, Islamic Republic of	448
Jordan	428	Indonesia	435
Iran, Islamic Republic of	422	Turkey	433
Indonesia	403	Tunisia	430
Chile	392	Chile	420
Philippines	345	Philippines	345
Morocco	337	Morocco	323
South Africa	275	South Africa	243
International average of 38 nations	**487**	**International average of 38 nations**	**488**

Average is significantly higher than the U.S. average

Average does not differ significantly from the U.S. average

Average is significantly lower than the U.S. average

[1] The shading of Finland may appear incorrect; however, statisically, its placement is correct.

[2] Designated LSS because only Latvian-speaking schools were tested, which represents 61 percent of the population.

[3] Lithuania tested the same cohort of students as other nations, but later in 1999, at the beginnng of the next school year.

Parentheses indicate nations not meeting international sampling standards or other guidelines. See NCES (2000) for details.

(Source: National Center for Education Statistics 2003c)

What It Means to Be Mathematically Literate

> 66 Learning with understanding is essential to mathematical literacy. 99

Learning with understanding is essential to mathematical literacy. When you hear the word *literacy*, you probably think of reading first. Not only does reading literacy mean being able to pronounce and decode words, it also means being able to comprehend and understand what one reads. Mathematical literacy is essentially the same thing—having procedural and computational skills as well as conceptual understanding.

The recent report *Adding It Up: Helping Children Learn Mathematics* (National Research Council 2002) states that mathematical proficiency has five interwoven and interdependent strands, as shown in the box at right.

NAEP data show that proficiency is developed unevenly among these strands. In many situations students are able to mimic rules and procedures demonstrated by their teacher; however, students learn these skills without much depth or understanding. Students are fairly successful in solving one-step problems but are much less successful when asked to solve more complex problems or apply their skills to new situations (Kouba and Wearne 2000; Wearne and Kouba 2000). Research suggests that one reason for this weakness is the separation between procedural and conceptual knowledge (Hiebert 1986; Hiebert and Carpenter 1992).

STRANDS OF MATHEMATICAL PROFICIENCY

1) Understanding (Conceptual Understanding): Comprehending mathematical concepts, operations, and relations—knowing what mathematical symbols, diagrams, and procedures mean.

2) Computing (Procedural Fluency): Carrying out mathematical procedures, such as adding, subtracting, multiplying, and dividing numbers, flexibly, accurately, efficiently, and appropriately.

3) Applying (Strategic Competence): Being able to formulate problems mathematically and to devise strategies for solving them using concepts and procedures appropriately.

4) Reasoning (Adaptive Reasoning): Using logic to explain and justify a solution to a problem or to extend from something known to something not yet known.

5) Engaging (Productive Disposition): Seeing mathematics as sensible, useful, and doable—if one works at it—and being willing to do the work.

Examples of High-Quality Mathematics Classrooms

Consider the following elementary, middle, and high school classroom scenarios:

Elementary School Classroom Example

This interchange occurred in an upper elementary class between a teacher and her students, who explained how they divided nine brownies equally among eight people (NCTM 2000, pp. 186–87):

Sarah: The first four we cut them in half. (*Jasmine divides squares in half on an overhead transparency.*)

Ms. Carter: Now as you explain, could you explain why you did it in half?

Sarah: Because when you put it in half, it becomes four … four … eight halves.

Ms. Carter: Eight halves. What does that mean if there are eight halves?

Sarah: Then each person gets a half.

Ms. Carter: Okay, that each person gets a half. (*Jasmine labels halves 1 through 8 for each of the eight people.*)

Sarah: Then there were five boxes [brownies] left. We put them in eighths.

Ms. Carter: Okay, so they divided them into eighths. Could you tell us why you chose eighths?

Sarah: It's easiest. Because then everyone will get … each person will get a half and (*addresses Jasmine*) … how many eighths?

Jasmine: (*Quietly*) Five-eighths.

Ms. Carter: I didn't know why you did it in eighths. That's the reason. I just wanted to know why you chose eighths.

Jasmine: We did eighths because then if we did eighths, each person would get each eighth, I mean one-eighth out of each brownie.

Ms. Carter: Okay, one-eighth out of each brownie. Can you just, you don't have to number, but just show us what you mean by that? I heard the words, but…

Jasmine: (*Shades in one-eighth of each of the five brownies that were divided into eighths.*) Person one would get this … (*points to one-eighth*)

Ms. Carter: Oh, out of each brownie.

Sarah: Out of each brownie, one person will get one-eighth.

Ms. Carter: One-eighth. Okay. So how much then did they get if they got their fair share?

Jasmine & Sarah: They got a half and five-eighths.

Ms. Carter: Do you want to write that down at the top, so I can see what you did? (*Jasmine writes 1/2 + 1/8 + 1/8 + 1/8 + 1/8 + 1/8 at the top of the overhead transparency.*)

Middle School Classroom Example

Students in a middle-grades classroom are working in pairs to determine the dimensions of a rectangle given the ratio of the length to width and the area (NCTM 2000, pp. 268–70):

The students began by working collaboratively in pairs to solve the following problem, adapted from Bennett, Maier, and Nelson (1998):

> A certain rectangle has length and width that are whole numbers of inches, and the ratio of its length to its width is 4 to 3. Its area is 300 square inches. What are its length and width?

As the students worked on the problem, the teacher circulated around the room, monitoring the work of the pairs and responding to their questions. She also noted different approaches that were used by the students and made decisions about which students she would ask to present solutions.

After most students had a chance to solve the problem, the teacher asked Lee and Randy to present their method. They proceeded to the overhead projector to explain their work. After briefly restating the problem, Lee indicated that 3 times 4 is equal to 12 and that they needed "a number that both 3 and 4 would go into." The teacher asked why they had multiplied 3 by 4. Randy replied that the ratio of the length to the width was given as "4 to 3" in the problem. Lee went on to say that they had determined that "3 goes into 15 five times and that 4 goes into 20 five times." Since 15 times 20 is equal to 300, the area of the given rectangle, they concluded that 15 inches and 20 inches were the width and length of the rectangle.

The teacher asked if there were questions for Lee or Randy. Echoing the teacher's query during the presentation of the solution, Tyronne said that he did not understand their solution, particularly where the 12 had come from and how they knew it would help solve the problem. Neither Lee nor Randy was able to explain why they had multiplied 3 by 4 or how the result was connected to their solution. The teacher then indicated that she also wondered how they had obtained the 15 and the 20. The boys reiterated that they had been looking for a number "that both 3 and 4 went into." In reply, Darryl asked how the boys had obtained the number 5. Lee and Randy responded that 5 was what "3 and 4 go into." At this point, Keisha said "Did you guys just guess and check?" Lee and Randy responded in unison, "Yeah!" Although Lee and Randy's final answer was correct and although it contained a kernel of good mathematical insight, their explanation of their solution method left other students confused.

To address the confusion generated by Lee and Randy, the teacher decided to solicit another solution. Because the teacher had seen Rachel and Keisha use a different method, she asked them to explain their approach. Keisha made a sketch of a rectangle, labeling the length 4 and the width 3. She explained that the 4 and 3 were not really the length and width of the rectangle but that the numbers helped remind her about the ratio. Then Rachel explained that she could imagine 12 squares inside the rectangle because 3 times 4 is equal to 12, and she drew lines to subdivide the rectangle accordingly. Next she explained that the area of the rectangle must be equally distributed in the 12 "inside" squares. Therefore, they divided 300 by 12 to determine that each square contains 25 square inches. At the teacher's suggestion, Rachel wrote a 25 in each square in the diagram to make this point clear. Keisha then explained that in order to find the length and width of the rectangle, they had to determine the length of the side of each small square. She argued that since the area of each square was 25 square inches, the side of each square was 5 inches. Then, referring to the diagram [she had drawn], she explained that the length of the rectangle was 20 inches, since it consisted of the sides of four squares. Similarly, the width was found to be 15 inches. To clarify their understanding of the solution, a few students asked questions, which were answered well by Keisha and Rachel.

High School Classroom Example

Students in a high school classroom are determining the short-est leash possible for a dog to guard a yard shaped like a right triangle (NCTM 2000, pp. 354–57):

The students in Mr. Robinson's tenth-grade mathematics class suspect they are in for some interesting problem solving when he starts class with this story: "I have a dilemma. As you may know, I have a faithful dog and a yard shaped like a right triangle. When I go away for short periods of time, I want Fido to guard the yard. Because I don't want him to get loose, I want to put him on a leash and secure the leash somewhere on the lot. I want to use the shortest leash possible, but wherever I secure the leash, I need to make sure the dog can reach every corner of the lot. Where should I secure the leash?"

After Mr. Robinson responds to the usual array of questions and comments (such as "Do you really have a dog?" "Only a math teacher would have a triangle-shaped lot—or notice that the lot was triangular!" "What type of dog is it?"), he asks the students to work in groups of three. All their usual tools, including compass, straight-edge, calculator, and computer with geometry software, are available. They are to come up with a plan to solve the problem.

Jennifer dives into the problem right away, saying, "Let's make a sketch using the computer." With her group's agreement, she produces the sketch in figure 7.36.

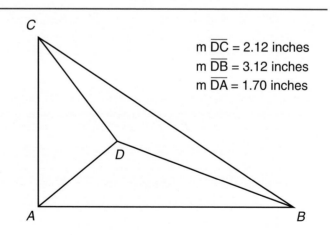

m $\overline{DC}$ = 2.12 inches
m $\overline{DB}$ = 3.12 inches
m $\overline{DA}$ = 1.70 inches

Fig. 7.36.

Jennifer's computer-drawn sketch of the "dog in the yard" problem

As Mr. Robinson circulates around the room, he observes each group long enough to monitor its progress. On his first pass, Jennifer's group seems to be experimenting somewhat randomly with dragging the point *D* to various places, but on his second pass, their work seems more systematic. To assess what members of the group understand, he asks how they are doing:

Mr. R: Joe, can you bring me up-to-date on the progress of your group?

Joe: We're trying to find out where to put the point.

Jeff: We don't want the point too close to the corners of the triangle.

Jennifer: I get it! We want all the lengths to be equal! They all work against each other.

Before moving on to work with other groups, Mr. Robinson works with the members of Jennifer's group on clarifying their ideas, using more-standard mathematical language, and checking with one another for shared understanding. Jennifer clarifies her idea, and the group decides that it seems reasonable. They set a goal of finding the position for *D* that results in the line segments *DA*, *DB*, and *DC* all being the same length. When Mr. Robinson returns, the group has concluded that point *D* has to be the midpoint of the hypotenuse, otherwise, they say, it could not be equidistant from *B* and *C*. (Mr. Robinson notes to himself that the group's conclusion is not adequately justified, but he decides not to intervene at this point; the work they will do later in creating a proof will ensure that they examine this reasoning.)

Mr. R: What else would you need to know?

Jeff: We're not sure yet whether *D* is the same distance from all three vertices.

Jennifer: It has to be! At least I think it is. It looks like it's the center of a circle.

Small-group conversations continue until several groups have made observations and conjectures similar to those made in Jennifer's group. Mr. Robinson pulls the class back together to discuss the problem. When the students converge on a conjecture, he writes it on the board as follows:

Conjecture: The midpoint of the hypotenuse of a right triangle is equidistant from the three vertices of the triangle.

He then asks the students to return to their groups and work toward providing either a proof or a counterexample. The groups continue to work on the problem, settling on proofs and selecting group members to present them on the overhead projector. As always, Mr. Robinson emphasizes the fact that there might be a number of different ways to prove the conjecture.

Remembering Mr. Robinson's mantra about placing the coordinate system to "make things eeeasy," one group places the coordinates as shown in figure 7.37a, yielding a common distance of $\sqrt{a^2 + b^2}$.

(Continued)

Alfonse, who is explaining this solution, proudly remarks that it reminds him of the Pythagorean theorem. Mr. Robinson builds on that observation, noting to the class that if the students drop a perpendicular from M to $\overline{AC}$, each of the two right triangles that result has legs of length a and b; thus the length of the hypotenuses, MC and MA, are indeed $\sqrt{a^2 + b^2}$.

Jennifer's group returns to her earlier comment about the three points A, B, and C being on a circle. After lengthy conversations with, and questions from, Mr. Robinson, that group produces a second proof based on the properties of inscribed angles (fig. 7.37b).

Fig. 7.37.

Diagrams corresponding to four proofs of the midpoint-of-hypotenuse theorem

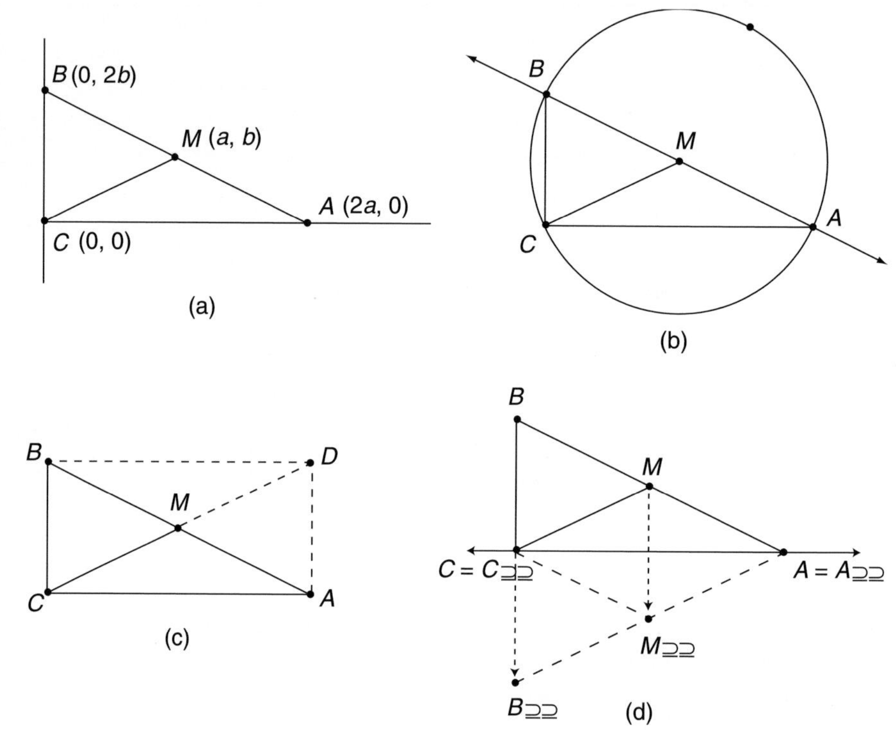

(a)

(b)

(c)

(d)

Elements of These Classrooms

These examples illustrate some of the elements of a classroom—whether it be elementary, middle, or high school—in which high-quality mathematics instruction and learning are taking place.

What Are Students Doing?

- Actively engaging in the learning process
- Using existing mathematical knowledge to make sense of the task
- Making connections among mathematical concepts
- Reasoning and making conjectures about the problem
- Communicating their mathematical thinking orally and in writing
- Listening and reacting to others' thinking and solutions to problems
- Using a variety of representations, such as pictures, tables, graphs, and words, for their mathematical thinking
- Using mathematical and technological tools, such as physical materials, calculators, and computers, along with textbooks and other instructional materials
- Building new mathematical knowledge through problem solving

What Is the Teacher Doing?

- Choosing "good" problems—ones that invite exploration of an important mathematical concept and allow students the chance to solidify and extend their knowledge
- Assessing students' understanding by listening to discussions and asking students to justify their responses
- Using questioning techniques to facilitate learning
- Encouraging students to explore multiple solutions
- Challenging students to think more deeply about the problems they are solving and to make connections with other ideas within mathematics
- Creating a variety of opportunities, such as group work and class discussions, for students to communicate mathematically
- Modeling appropriate mathematical language and a disposition for solving challenging mathematical problems

Although the content changes as students progress through the grades, the statements mentioned above for teachers and students are common characteristics that you should see in any mathematics classroom. In all these scenarios, a climate has been created that supports mathematical thinking and communication. Students are accustomed to explaining their ideas and questioning solutions that might not make sense to them. Students are not afraid to take risks and know that it is acceptable to struggle with some ideas and to make mistakes. The teacher responds in ways that keep the focus on thinking and reasoning rather than only on getting the right answer. Incorrect answers and ideas are not simply judged wrong. Teachers help students identify parts of their thinking that may be correct, sometimes leading students to new ideas and solutions that are correct.

> **" A climate has been created that supports mathematical thinking and communication. "**

In all three of the foregoing classroom examples, the very important role that the teacher plays is evident. While giving students plenty of opportunities to think for themselves and come up with their own solutions, the teacher is crucial to facilitating and guiding the mathematical tasks and conversations. Achieving this kind of classroom requires much skill and judgment on the part of the teacher, as well as a solid understanding of the mathematics content. It also requires schoolwide changes and support for classroom teachers. The following section discusses six guiding principles that you and your teachers can use to improve mathematics education in your school.

The Six NCTM Principles: The Foundation of a High-Quality Mathematics Program

Principles and Standards for School Mathematics (NCTM 2000) proposes six principles—Equity, Curriculum, Teaching, Learning, Assessment, and Technology—as a foundation for high-quality mathematics programs.

The chart below and the subsequent pages summarize the main ideas for each principle. The section titled "Putting the NCTM Principles into Action in Your School" focuses on specific actions that you can take to support these principles in your school.

EQUITY	CURRICULUM
Excellence in mathematics education requires equity—high expectations and strong support for all students.	*A curriculum is more than a collection of activities: it must be coherent, focused on important mathematics, and well articulated across the grades.*
◆ Equity requires high expectations and worthwhile opportunities for all.	◆ A mathematics curriculum should be coherent.
◆ Equity requires accommodating differences to help everyone learn mathematics.	◆ A mathematics curriculum should focus on important mathematics.
◆ Equity requires resources and support for all classrooms and students.	◆ A mathematics curriculum should be well articulated across the grades.

TEACHING

Effective mathematics teaching requires understanding what students know and need to learn and then challenging and supporting them to learn it well.

◆ Effective teaching requires knowing and understanding mathematics, students as learners, and pedagogical strategies.

◆ Effective teaching requires a challenging and supportive classroom learning environment.

◆ Effective teaching requires continually seeking improvement.

LEARNING

Students must learn mathematics with understanding, actively building new knowledge from experience and prior knowledge.

◆ Learning mathematics with understanding is essential.

◆ Students can learn mathematics with understanding.

ASSESSMENT

Assessment should support the learning of important mathematics and furnish useful information to both teachers and students.

◆ Assessment should enhance students' learning.

◆ Assessment is a valuable tool for making instructional decisions.

TECHNOLOGY

Technology is essential in teaching and learning mathematics; it influences the mathematics that is taught and enhances students' learning.

◆ Technology enhances mathematics learning.

◆ Technology supports effective mathematics teaching.

◆ Technology influences what mathematics is taught.

> **" Too many students are victims of low expectations in mathematics. "**

Equity

Research indicates that all students can learn mathematics when they are supported by, and have access to, high-quality mathematics instruction (Campbell 1995; Griffin, Case, and Siegler 1994; Knapp et al. 1995; Silver and Stein 1996). Traditionally, mathematics has been seen as something that only a select few could master. High-quality mathematics education is not just for those who want to study mathematics and science in college. It is for everyone. Too many students—especially students who are poor, not native speakers of English, disabled, female, or members of racial-minority groups—are victims of low expectations in mathematics. For example, tracking has consistently disadvantaged groups of students by relegating them to mathematics classes that concentrate on remediation or do not offer significant mathematical substance.

Equity Misunderstandings

Equity does not mean that all students will have identical instruction. It does mean that reasonable and appropriate accommodations will be made so that all students have the opportunity to experience a common foundation of challenging mathematics. Certainly, students with special interest and exceptional talent in mathematics may need enrichment programs or additional resources to challenge and engage them. Likewise, some students may need further assistance to meet high mathematics expectations.

Curriculum

Curriculum in the United States has often been criticized as being "a mile wide and an inch deep." Students study many topics each year—often reviewing topics that were covered in previous years—and very little depth is added each time the topic is addressed. The higher performing countries, in contrast, tend to select a few fundamental topics each year and spend a considerable amount of time developing those topics (Fuson, Stigler, and Bartsch 1988; McKnight et al. 1987; McKnight and Schmidt 1998; Peak 1996). Mathematics is a highly interconnected and cumulative subject. Therefore, the mathematics curriculum needs to introduce ideas in ways that build on one another. Students should have opportunities to learn increasingly more sophisticated mathematical ideas as they progress through the grades.

The NCTM Standards for Prekindergarten through Grade 12

Principles and Standards for School Mathematics (NCTM 2000) reflects current thinking and research on mathematics teaching and learning and should be used as a resource for curriculum decisions. The Standards describe the mathematical understanding, knowledge, and skills that all students should have the opportunity to learn in prekindergarden through grade 12.

The Content Standards—Number and Operations, Algebra, Geometry, Measurement, and Data Analysis and Probability—describe the mathematical content students should learn. The Process Standards—Problem Solving, Reasoning and Proof, Communication, Connections, and Representation—highlight ways of acquiring and using content knowledge. Together, these ten Standards define the basic mathematics that all students should have the opportunity to learn.

Mathematics for All		Dumbing Down
Cooperative Grouping	**DOES NOT IMPLY** ➔	Lack of Accountability
Heterogeneous Grouping		Removing the Gifted Track

The Table of Standards and Expectations at the end of this guide highlights the goals and expectations described in *Principles and Standards* for each of the four grade bands (pre-K–2, 3–5, 6–8, and 9–12). As can be seen from this table, the curriculum should be set up so that students build a successively deeper and more refined understanding of each of the Standards as they move through the grades. For example, let us look at the Geometry Standard and how students' knowledge of two-dimensional shapes might grow from prekindergarten through grade 2. "In grades K–2 students typically explore similarities and differences among two-dimensional shapes. In grades 3–5 they can identify characteristics of various quadrilaterals. In grades 6–8 they may examine and make generalizations about properties of particular quadrilaterals. In grades 9–12 they may develop logical arguments to justify conjectures about particular polygons" (NCTM 2000, p. 16). Articulation across the grades is crucial so that teachers at each grade understand the mathematics that has been studied by students in the previous grade and what is to be the focus in successive grades.

Teaching

We know that teaching mathematics well is a complex endeavor and that no easy recipes exist. Teachers must have both content and pedagogical knowledge. For teachers to help students engage in rich and challenging mathematics, they must have a deep understanding of the mathematics they teach. In addition, teachers must understand the curriculum goals and the ideas that are central to their grade band, as well as the grades before and after it. The pedagogical knowledge needed by teachers includes knowing how students learn mathematics, how to build on students' existing knowledge, how ideas can be presented to students so they gain a deep understanding, and how students' learning can be assessed.

> **" Teachers must have both content and pedagogical knowledge. "**

Professional Standards for Teaching Mathematics

Professional Standards for Teaching Mathematics (NCTM 1991) presented six standards for teaching mathematics. They address the following considerations:

1. **Worthwhile Mathematical Tasks**

   Teachers should choose and develop tasks that are likely to promote the development of students' understandings of concepts and procedures and that foster students' ability to solve problems and communicate mathematically.

2. **The Teacher's Role in Discourse**

   Teachers play a crucial role in supporting discourse in ways that facilitate learning without taking over the process of thinking for students.

3. **The Students' Role in Discourse**

   Students should be accustomed to making conjectures, asking questions, and proposing approaches and solutions to problems.

4. **Tools for Enhancing Discourse**

   Various tools for communicating about mathematics should be encouraged, for example, concrete materials, diagrams, graphs, tables, calculators, computers, and other technological tools.

5. **The Learning Environment**

   Essential dimensions of a learning environment in which serious mathematical thinking can take place include a genuine respect for others' ideas, a valuing of reason and sense-making, pacing and timing that allow students to puzzle and think, and forging a social and intellectual community.

6. **The Analysis of Teaching and Learning**

   Teachers should monitor students' learning on a regular basis for the purpose of modifying and adjusting instruction.

Despite all we have learned about teaching and learning mathematics, teaching practices in the United States have remained virtually unchanged over the last century (National Advisory Committee on Mathematics Education 1975; Fey 1979; Stigler and Hiebert 1997). This teaching style, illustrated in the TIMSS videotape study of eighth-grade mathematics classes in the United States, is typically as follows:

1. Review of previous material and homework
2. A problem demonstrated by the teacher
3. Drill on low-level procedures
4. Supervised seatwork by students, often in isolation
5. Checking of seatwork problems
6. Assignment of homework

This teaching style has sometimes been termed *recitation* (Hoetker and Ahlbrand 1969; Tharp and Gallimore 1998). New material is presented to students largely by telling them or demonstrating procedures to them. Little connection is made between the procedures and the mathematical concepts. Students practice similar problems to those demonstrated by the teacher. For example, the teacher might simply state that the area of a right triangle is "1/2 times the base height" rather than let students try to determine this formula on the basis of their knowledge that the area of a parallelogram is "base times height." This approach is in contrast with the teaching style in some of the top-performing countries, including Japan. For example, the TIMSS videotape study revealed that in Japan, only 17 percent of mathematical concepts were stated by the teacher rather than developed in a more reasoned approach, whereas this kind of practice occurred for 78 percent of the concepts presented in U.S. lessons (Stigler and Hiebert 1999).

Learning

Learning mathematics with understanding is the cornerstone of the NCTM vision for mathematics education. Unfortunately, learning without understanding has long been a common outcome of school mathematics instruction (e.g., Brownell [1947]; Skemp [1976]; Hiebert and Carpenter [1992]). Research supports the fact that that conceptual understanding must be present for students to become proficient in mathematics. Having factual and procedural knowledge is also important; however, memorizing facts and procedures in isolation results in students' failure to know when to use and apply those skills to solve problems (Bransford, Brown, and Cocking 1999). Learning mathematics with understanding requires active engagement in meaningful tasks and experiences designed to deepen and connect students' existing knowledge with new knowledge.

Summary of Students and Learning during the Four Grade Bands

Prekindergarten through Grade 2. The foundation for mathematical development is established early. Children in prekindergarten through grade 2 bring a substantial amount of informal knowledge to school. They enter school curious and eager to learn more about numbers and mathematical objects. They are learning to understand patterns and measurement and develop a solid understanding of the numeration system. Although their levels of mathematical understanding when entering school vary, teachers and others should realize that "not knowing" is often a reflection of a lack of opportunity to learn and not an inability to learn. Early assessments should be used not to sort children but to gain information for teaching and for potential early interventions. Children at these grades also tend to know more than they can express in writing, so a variety of assessment methods, such as interviews and teacher observations, should be used.

Grades 3 through 5. Most students enter these grades with a continued interest in mathematics. However, students begin to lose interest in mathematics when it simply becomes a process of mimicking and memorizing. Instruction should continue to be active and intellectually stimulating. Three central ideas that are developing during these grades are multiplicative reasoning, equivalence, and computational fluency. Students should develop computational fluency with whole numbers on the basis of strong conceptual understanding.

Grades 6 through 8. Students reach a significant turning point in their lives during the middle grades. Students in these grades are experiencing physical, emotional, and intellectual changes. During these grades many students form conclusions about their mathematical abilities and about their interest in, and attitude toward, mathematics. Unfortunately, too many students begin to enjoy mathematics less during these years and think negatively about their own mathematical abilities. Middle-grades students will continue to appreciate and enjoy mathematics if they find both challenge and support in the mathematics classroom. Proportionality, algebraic and geometric thinking, and rational numbers are central themes throughout the middle grades. Middle school mathematics classes should be designed so that all middle-grades students experience a rich and integrated treatment of mathematics content throughout grades 6, 7, and 8.

Grades 9 through 12. The high school years are also a major time of transition. Students in these grades are continuing to develop in many ways. They are becoming both autonomous and also able to work better with others. They are able to reflect more and are developing personal and intellectual abilities that will prepare them for the workplace and postsecondary education. Because students' interests and aspirations may change during and after high school, their mathematics education should prepare them for a variety of career and education options. All students should study mathematics in each of the four years that they are enrolled in high school. They should experience the interplay of algebra, geometry, statistics, probability, and discrete mathematics and should develop deeper understandings of the fundamental mathematical concepts of function and relation, invariance, and transformation.

Assessment

Assessment is an important tool in any mathematics program. Assessment has traditionally been thought of as a test at the end of instruction to measure students' attainment of specified goals after instruction. Although this outcome is one purpose of assessment, assessments also serve other purposes, such as guiding teachers' instructional practice. Research supports the precept that students' learning is improved when assessment is an integral part of ongoing classroom activities (Black and William 1998). Good assessments provide useful information about students' learning, as well as for instructional practice.

The chart below divides the purpose of assessments in mathematics into four broad categories: (1) monitoring students' progress, (2) making instructional decisions, (3) evaluating students' achievement, and (4) evaluating programs; and it lists four actions that result from the use of assessment data in conjunction with each of these: (1) promote growth, (2) improve instruction, (3) recognize accomplishment, and (4) modify program.

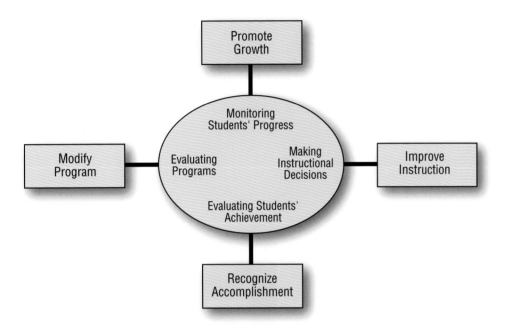

Source: *Assessment Standards for School Mathematics* (NCTM 1995, p. 25)

Daily Classroom Assessment

Assessment should be part of the mathematics teacher's daily practice. Assessments before and during instruction allow teachers to make appropriate decisions about such considerations as reviewing material, reteaching a difficult concept, or providing additional or more challenging material for students who are struggling or need enrichment. Teachers can use many assessment techniques, including open-ended questions, constructive-response tasks, selected-response items, performance tasks, observations, discussions, journals, and portfolios. Some of these techniques are more useful for specific goals. For example, a quiz or test with simple constructive-response tasks and selective-response items might be used to assess students' ability to execute procedures. More-complex constructive-response items and performance tasks offer a better assessment of students' ability to apply mathematics. Observations of students working in groups and of classroom discussions can provide a teacher with useful information regarding students' thinking. Journals and portfolios can be used to assess students' thinking and reasoning over time (NCTM 2000). Students show what they know and can do in different ways. Relying on only one type of assessment may give an incomplete picture of students' actual understanding and performance.

Understanding Large-Scale Assessments

Although large-scale tests can provide useful data, they often limit the scope of what is being tested to what fits the test format—usually a multiple-choice format. These types of tests do not always provide an accurate view of students' conceptual understanding or their problem-solving abilities.

Students' scores on these tests can also be deceiving, depending on how the scores are reported. Often, overall mean scores from a norm-referenced test or a district-by-district comparison from a statewide assessment are used to make important decisions about the success of a mathematics program and the extent of students' learning. Using mean scores alone does not consider enough information. For example, average percentiles do not convey information about the dispersion of scores being averaged. How many students scored higher? How much lower did other students score? Not always evident in these scores are other important questions, such as, Are boys and girls and students of all ethnicities showing the same progress? What progress are students making in each of the curriculum components? (NCTM 1995)

The current high-stakes testing environment can also force many teachers to focus so narrowly on one particular test that scores may rise without actual improvement in the broader set of skills that today's students need. You as an administrator need to understand how to analyze these scores and know what these scores can and cannot reveal. More important, you should promote daily assessment practices in the mathematics classroom and rely on those assessments in conjunction with other larger-scale assessments to gain a more accurate picture of students' achievement.

concepts and skills. The depth of problem solving that students can pursue with proper use of technology is astounding. Technology can provide powerful visual images of mathematical ideas, can help organize and analyze data, and can efficiently and accurately perform mathematical computations so that students can focus on conceptual understanding and higher-order thinking. Preparing all students to use the technological tools that are available to them will also be essential for the workplace and for success in our information-based and technologically based society.

Technology

Students should have access to a full range of technological tools. Students can learn more mathematics more deeply with the appropriate uses of technology (Dunham and Dick 1994; Sheets 1993; Boers-van Oosterum 1990; Rojano 1996; Groves 1994). The teacher plays the central role of ensuring that technology is used appropriately to support students' learning of mathematics, not as a crutch or a replacement for the mastery of basic

Putting the NCTM Principles into Action in Your School

The six NCTM Principles described in the previous section—Equity, Curriculum, Teaching, Learning, Assessment, and Technology—are the basic precepts that are fundamental to a high-quality mathematics program. Thinking about these principles and ways they can be supported, both schoolwide by administrators and at the classroom level by individual teachers, is an important step in improving mathematics education in your school.

Actions Administrators Can Take to Support the Six NCTM Principles

The role of teachers is central to a high-quality mathematics program. However, many decisions made outside the classroom affect the teacher's ability to carry out this vision. As an administrator, you serve an important role in this effort by shaping the

instructional mission in your school, providing for the professional development of teachers, designing and implementing policies, and allocating resources.

Again, consider the six principles as you engage in decision making that affects school mathematics. For these principles to be achieved at the classroom level, the school environment must be structured to encourage and support them. The following are some actions you can take to support each of the principles.

> " Thinking about [the six] principles is an important step in improving mathematics education in your school. "

OVERALL QUESTIONS TO REFLECT ON

Equity:
How can all students have access to high-quality mathematics education?

Curriculum:
Are good instructional materials chosen, used, and accepted?

Teaching:
How can teachers learn what they need to know?

Learning:
Do all students have time and opportunities to learn?

Assessment:
Are assessments aligned with instructional goals?

Technology:
Is technology supporting learning?

Principle	Action
Equity *Excellence in mathematics education requires equity—high expectations and strong support for all students.*	◆ Create a school climate built on the expectation of high achievement by all students. ◆ Energize teachers and students in ways that challenge current expectations. ◆ Evaluate the processes for placing students in mathematics classes to ensure that groups of students are not being excluded from a challenging mathematics program.
Curriculum *A curriculum is more than a collection of activities; it must be coherent, focused on important mathematics, and well articulated across the grades*	◆ For selecting instructional materials, establish processes that involve teachers and teacher-leaders and that provide for a careful analysis of these materials. ◆ Set up a program in your school to allow for articulation across the grades as well as from elementary school to middle school to high school. ◆ Help families understand the goals of the curriculum. ◆ Provide access to resources and instructional materials that support the curriculum.
Teaching *Effective mathematics teaching requires understanding what students know and need to learn and then challenging and supporting them to learn it well.*	◆ Support sustained and ongoing professional development that is tied to the curriculum and that increases teachers' mathematics and pedagogical knowledge. ◆ Support teachers in self-evaluation and in analyzing, evaluating, and improving their teaching with colleagues and supervisors.

Principle	Action
Teaching—Continued	◆ Arrange teachers' work schedule so that meaningful collaboration with colleagues is part of the school day.
	◆ Establish teacher-leaders or mathematics specialists who can mentor and support teachers.
	◆ Spend time observing mathematics classrooms.
	◆ Recruit qualified teachers of mathematics.
	◆ Make teaching assignments on the basis of the qualifications of teachers.
	◆ Promote and support your teachers' attendance at professional conferences.
	◆ Attend professional development sessions designed to help administrators understand the goals of mathematics instruction.

Learning
 	
Students must learn mathematics with understanding, actively building new knowledge from experience and prior knowledge.
	◆ Ensure that sufficient time is allocated for mathematics instruction.
	  — Elementary school students should study mathematics for at least an hour a day under the guidance of teachers who enjoy mathematics and are prepared to teach it well.
	  — Middle school and high school students should study the equivalent of a full year of mathematics in each grade.

	◆ Promote the importance of learning mathematics with understanding to teachers and families.
	◆ Develop a plan to identify and support students who are struggling in mathematics.

(Continued)

Principle	Action
Assessment *Assessment should support the learning of important mathematics and furnish useful information to both teachers and students.*	◆ Ensure that assessments are aligned with the curriculum. ◆ Examine the impact of high-stakes assessments on the instructional climate in schools. ◆ Ensure that decisions about placing students in mathematics classes and evaluations of teachers' effectiveness are not based on a single test. ◆ Ensure that teachers are using a variety of classroom assessment methods that measure conceptual understanding along with factual and procedural understanding. ◆ Ensure that teachers rely on daily formative assessment to plan and evaluate instruction.
Technology *Technology is essential in teaching and learning mathematics; it influences the mathematics that is taught and enhances students' learning.*	◆ Ensure that technology is being used to enhance learning. ◆ Ensure equitable access to technology. ◆ Develop a coherent and comprehensive plan to embed technology into the mathematics curriculum. ◆ Support professional development to help teachers implement technology in mathematics instruction.

Observing and Evaluating a Mathematics Classroom

What should you look for when observing a mathematics classroom? The previous section titled "Examples of High-Quality Mathematics Classrooms" presented elementary, middle, and high school scenarios to illustrate many of the things you should observe in the classroom.

You should also see evidence of the six NCTM Principles—Equity, Curriculum, Teaching, Learning, Assessment, and Technology—when observing a mathematics classroom. The following are some questions to think about during a classroom observation. The answers to these questions may not always be evident during a particular observation or may not be easily observable. Multiple observations and one-to-one discussions with the teacher are important to obtain a complete picture of a teacher's instructional practice. Encouraging the teacher to reflect on some of these questions after the observation is also helpful. These observations and discussions not only provide you with information that cannot be observed but also support the teacher's professional growth and help him or her consider the six Principles while planning and delivering mathematics instruction.

Equity

- What strategies and accommodations did the teacher use so that all students experienced a common foundation of challenging mathematics?
- What words and actions did the teacher communicate to show high expectations for all students?
- In what ways did the teacher create a climate that helped all students develop confidence?

Curriculum

- Did the teacher effectively organize the lesson so that the important mathematical ideas formed an integrated whole?
- Were the fundamental ideas presented to students in a variety of contexts?
- Did the teacher choose tasks and activities that supported the curriculum?

Teaching and Learning

- Did the teacher frequently introduce concepts and skills through problem-solving or reasoning experiences rather than simply tell students or demonstrate procedures to them?
- What evidence did you see that the teacher built on students' existing knowledge and understanding to solve new problems?
- Did the tasks and problems presented by the teacher lend themselves to students' discovery of an important mathematical concept or further develop their understanding of concepts and procedures?
- Did the teacher give students enough time to think and reason for themselves?
- Did the teacher use such statements as "Explain" and "Why" to encourage mathematical communication?
- Did the teacher create a variety of opportunities for students to communicate their mathematical thinking orally and in writing?
- Did students seem to be accustomed to making conjectures, explaining their ideas, and questioning solutions that do not make sense to them?
- Did the teacher encourage and accept multiple approaches to, and representations of, a problem?
- Did the teacher respond in ways that kept the focus on thinking and reasoning rather than only on getting the right answer?
- Did the teacher help students identify the parts of their thinking that may be correct?
- In what ways did the teacher make connections between the mathematics being studied and other ideas in mathematics, other subject areas, or the real world?

- What assessment techniques did the teacher use to ensure that both conceptual and procedural knowledge are assessed, as well as students' ability to apply mathematics in complex and new situations?
- In what ways did the teacher use daily assessment practices to guide instructional practice and to make decisions about the needs of individual students?

Technology

- If technology was used, what evidence did you see that the technology fostered a deeper understanding of the mathematics?
- If technology was not used, could it have been used to make some ideas more accessible to students?
- Did the teacher help students determine when paper and pencil, technology, or mental techniques are most appropriate?

Developing and Supporting Professional Development

Perhaps the best way to have a positive impact on mathematics achievement for all students in your school is by making a substantial investment in professional development. National and international studies of students' achievement highlight the importance of staffing schools with teachers who are properly prepared to teach mathematics. Preservice preparation is the foundation for mathematics teaching, but it gives teachers only a small part of what they will need to know and understand throughout their careers. Despite the many myths that teaching is something one is "born with," the ability to teach is learned and refined over time. Teachers, like all other professionals, need ongoing and sustained professional development opportunities throughout their careers. The current practice of offering occasional workshops and in-service days is not enough.

ELEMENTS OF EFFECTIVE PROFESSIONAL DEVELOPMENT PROGRAMS

◆ Develops teachers' knowledge of—
- mathematics content,
- students and how they learn mathematics,
- effective instructional and assessment practices

◆ Models examples of high-quality mathematics teaching and learning

◆ Allows teachers to reflect on their practice and student learning in their classroom

◆ Allows teachers to collaborate and share experience with colleagues

◆ Connects to a comprehensive long-term plan that includes student achievement

Knowledge Needed by Mathematics Teachers

Teachers of mathematics need three core areas of knowledge (National Research Council 2001):

- Knowledge of mathematics
- Knowledge of students
- Knowledge of instructional and assessment practices

Professional development programs should address these three areas. One of the weakest areas for teachers in the United States, especially those in elementary and middle grades, is knowledge of mathematics. This requirement is more than knowing mathematics for oneself. It includes the ability to understand the conceptual basis for the mathematical knowledge and the ability to explain and teach it in ways that make sense to students.

The answer is not simply to require teachers to take more standard college mathematics courses. The topics typically taught at this level are often far removed from the core content of the K–12 curriculum and do not emphasize the conceptual underpinnings needed by teachers in these grades (National Research Council 2001). Although improving teachers' mathematical knowledge should be a primary component of any professional development program, professional development should connect the knowledge of mathematics with the knowledge of students and instructional practices.

Regular Teacher Collaboration—Building a Professional Learning Community

Research indicates that teachers are better able to help their students learn mathematics when they have opportunities to work together to improve their practice, time for personal reflection, and strong support from colleagues and other qualified professionals (see, e.g., Brown and Smith [1997]; Putnam and Borko [2000]; Smith [2000]). Unfortunately, many teachers work in relative isolation and meet with other teachers only to discuss administrative details.

Teachers should be given opportunities to collaborate with colleagues on a regular basis. Unfortunately, teachers are not often given enough time during the school day for this type of professional development. The typical structures of U.S. teachers' workdays often inhibit this kind of professional community. In Japan and China, however, the workdays of mathematics teachers include time for such collaboration (Ma 1999; Stigler and Hiebert 1999). As an administrator, you play a role in determining how such learning communities can be established in your school. One idea is to establish school-level mathematics teacher-leaders or specialists who are strong leaders and have expertise in mathematics teaching and learning. These leaders can work directly with teachers at the classroom level and can also assist in designing schoolwide professional development plans.

> **" Teachers should be given opportunities to collaborate with colleagues on a regular basis. "**

School Leadership That Supports Teachers' Professional Growth

Studies from the National Center for Improving Student Learning and Achievement in Mathematics and Science (NCISLA) have revealed the need for new conceptions of school leadership to support instructional reform. In these studies, teachers were successful in changing their instructional practices when leadership was distributed throughout the organization and teachers collaborated with administrators on decision making. Adam Gamoran, a sociologist from NCISLA, notes that this approach "requires administrators to rethink their roles from the director of an activity to a facilitator—someone who draws connections, someone who makes linkages, someone who supports the initiatives that teachers are taking" (Gamoran, Anderson, Quiroz, Secada, Williams, and Ashmann, forthcoming). As a result, teacher-leaders emerge and new human and social resources are generated.

Supporting New Teachers

Turnover in the nation's teaching force is high, especially with new teachers. A significant percent of new teachers leave the profession within their first three years of service. Beginning teachers are often hired at the last moment, left isolated in their classrooms, and given little support. The beginning years of teaching are very challenging and often overwhelming for many teachers. Along with professional development, you can help build these teachers' confidence and interest in teaching in the following ways:

- Provide new teachers with frequent interactions with master teachers, including class observations and teacher critiques.
- Develop a mentoring program in your school for new teachers.
- Develop policies that ensure that new teachers do not inherit the most demanding schedules and most challenging students.
- Develop policies that limit new teachers' extracurricular responsibilities.

Identifying Instructional Materials

Teachers and students need instructional materials that support high-quality teaching and learning. As an administrator, you can ensure a thorough process for analyzing and selecting instructional materials.

Who Should Be Involved in Identifying Instructional Materials?

First, you should ensure that the teachers who will be using these materials are involved in the identification process. Also, teacher-leaders and others who have expertise in the mathematical content at the particular grade level should be involved.

How Do You Analyze Instructional Materials?

Those who have expertise in the mathematical content should pay close attention to the sequence, timing, developmental appropriateness, and complexity of mathematical tasks described in these materials. You will also find that asking the following questions is helpful when evaluating instructional materials:

- Do the teaching materials ask students to perform at high cognitive levels?
- Do the materials help teachers understand the content for themselves and foster a better understanding of teaching and learning mathematics?
- Do the materials integrate assessment into the teaching and learning processes?
- Do the materials lead to conceptual development over time, that is, do they build on students' content knowledge from grade to grade?

Why Is Family Involvement Important?

Forming strong relationships with families is important to a successful mathematics program. Research supports the conclusion that parents' attitudes toward their children's education, and their involvement in it, have a significant impact on classroom success.

Parents can help their children have a good attitude about mathematics. Adults frequently make comments such as "I can't do math" or "I don't like math." By contrast, a parent is much less likely to say, "I can't read." When a parent—the child's role model—says that he or she does not like or cannot do mathematics, the statement is a signal that the same sentiment is OK on the child's part. Parents need to be educated about how their feelings about mathematics can affect their children's thinking about mathematics and about themselves as mathematicians. They need to understand that mathematical literacy is just as important as reading literacy.

Parents can also serve as advocates for the mathematics program. For parents to do so, they must understand the goals of the mathematics program and the reasons these goals are important. Collaborating with parents and inviting them to participate in efforts to improve the mathematics program are essential for a successful mathematics program for all.

> **66** **Parents need to understand that mathematical literacy is just as important as reading literacy. 99**

What Information and Resources Should Be Communicated to Families?

The mathematics classroom today may look very different from the classrooms that parents experienced when they were in school. Some parents may even feel uncomfortable or have misconceptions about the mathematics their children are learning. Therefore, an important role for the administrator is to provide parents with information that will help them understand the mathematics program and ways they can contribute to the success of their children's learning.

1. Inform parents of the goals of the mathematics programs, what their children will be learning, and why these goals are important.

2. Provide information on how parents can support their children's learning in school and at home.

3. Offer hands-on experiences for parents so that they understand and appreciate the mathematics their children are learning.

4. Communicate to parents the expectation that all students can be successful in mathematics.

Ways to Communicate with Families

1. Provide information about the mathematics program in parent newsletters and on the school World Wide Web site.

2. Host a schoolwide "family math night" in which parents participate in mathematics activities with their children.

3. Inform parents of the mathematics program during back-to-school nights and parent-teacher conferences.

4. Encourage parents to volunteer in the mathematics classroom.

The NCTM Figure This! Program

Through a grant from the National Science Foundation, NCTM in conjunction with Widmeyer Communications and the National Action Council for Minorities in Engineering developed a program to involve families of middle-grades students in mathematics. In addition to mathematics-challenge problems, the program developed family-support materials with ideas and information to give to parents. All these materials can be found on the World Wide Web at www.figurethis.org.

Communicating with Others in the School System and Community

Making the vision of mathematics teaching and learning described in the NCTM *Standards* documents a reality requires understanding and support from a variety of other stakeholders in the school system and community, including school board members, district and state administrators, business and community leaders, college and university faculty, leaders of professional organizations, and policymakers in government. You are in the position to reach out to some of these audiences and build support for high-quality mathematics programs. Just as you communicate to parents, also help these parties understand the goals and priorities of mathematics education today, as well as know the kind of support needed from them.

Conclusion

The following excerpt from *Principles and Standards for School Mathematics* (NCTM 2000, pp. 367–68) effectively summarizes in a broader context what is needed to ensure a high-quality mathematics education for all students. As an administrator, you can play a major role in helping make many of these ideas a reality in your school.

Imagine that all mathematics teachers continue to learn new mathematics content and keep current on education research. They collaborate on problems of mathematics teaching and regularly visit one another's classrooms to learn from, and critique, colleagues' teaching. In every school and district, mathematics teacher-leaders are available, serving as expert mentors to their colleagues, recommending resources, orchestrating interaction among teachers, and advising administrators. Education administrators and policymakers at all levels understand the nature of mathematical thinking and learning, help create professional and instructional climates that support students' and teachers' growth, understand the importance of mathematics learning, and provide the time and resources for teachers to teach and students to learn mathematics well. Institutions of higher learning collaborate with schools to study mathematics education and to improve teacher preparation and professional development. Professional mathematicians take an interest in, and contribute constructively to, setting the content goals for mathematics in grades K–12 and for developing teachers' mathematical knowledge. Professional organizations, such as the National Council of Teachers of Mathematics, provide leadership, resources, and professional development opportunities to improve mathematics education. And families, politicians, business and community leaders, and other stakeholders in the system are informed about education issues and serve as valuable resources for schools and children.

Q Are the traditional basics still important?

A Absolutely! A major goal in the early grades—prekindergarten through grade 5—is the development of computational fluency with whole numbers. In grades 6–8, students should acquire computational fluency with fractions, decimals, and integers. Basic computation skills continue to be needed in grades 9–12, in which students should develop fluency in operations with real numbers, vectors, and matrices.

Even in complex calculations in which calculators or technology are being used, students need to understand the mathematics. In today's world, students' basic arithmetic skills must include the ability to choose what numbers to use and what operation is appropriate for carrying out the computation, to decide whether the results make sense, and then to make a decision about what to do next.

Q What mathematics beyond the traditional basics should all students learn?

A To succeed in today's and tomorrow's world, students need more than the traditional basics. The basics are changing. Arithmetic skills, although important, are no longer enough. To succeed in tomorrow's world, students must have an understanding of the basic concepts of algebra, geometry, measurement, and data analysis and probability. All these topics describe a new definition of basic mathematics skills.

Q How should students be grouped?

A Historically, students have often been grouped according to their perceived mathematical abilities. The students in the "higher ability" classes tended to experience mathematics content that challenged and enriched them, whereas the "lower ability" groups were placed in classes with lower expectations and a focus on remediation. Also, those students identified as lower achieving tended to stay in that group throughout their schooling, even though they may have been able to move up and do more challenging work. Schools face difficult decisions about grouping. Structures that exclude certain groups of students from

a challenging, comprehensive mathematics program should be dismantled. An important point to remember is that some higher-achieving countries do not separate high achievers from low achievers. Students can effectively learn mathematics in heterogeneous groups if structures are developed to provide appropriate and differentiated support for a range of students.

Q What is the role of practice or drill in instruction?

A Practice is important, but not without understanding. Once students understand a computational procedure, practice will help them become confident and competent in using it. However, practice without understanding may be detrimental to students' understanding. When students mimic a procedure without understanding, they often have difficulty going back later on and building understanding. Drilling students on facts and procedures without emphasizing understanding also leads students to think that memorization is the key to mathematical power and does not help them understand that mathematics is about thinking and reasoning. Procedural skills should always be developed and assessed alongside conceptual understanding.

Q What are the appropriate uses of manipulatives in the mathematics classroom?

A Mathematics achievement at all grades is enhanced through concrete instructional materials and active participation. Manipulatives are physical objects, such as base-ten blocks, algebra tiles, geometric solids, coins, and tongue depressors, that can make abstract ideas and symbols more meaningful and understandable to students. However, the manipulatives by themselves do not result in learning. Students will not automatically make connections between the manipulatives and the concepts being explored. Manipulatives, just like technology and any other mathematical tool, need to support students' conceptual understanding. Teachers need to ensure that these tools are being used to enhance learning and should help students see the connections between these tools and the mathematical concepts. The teacher should also help students move from using the manipulatives to numerically and symbolically solving problems. Teachers will benefit from professional development opportunities focused on effectively incorporating manipulatives into their mathematics instruction.

Q Will calculators and other technology hurt students' computational skills?

A Using technology does not mean an abandonment of pencil-and-paper, mental, and other computational strategies. It means that students learn to use technology in appropriate settings and know when using technology makes sense. Studies even show that no computational ability is lost when technology is purposefully integrated into teaching. Data from the 1996 National Assessment of Educational Progress (NAEP) showed that fourth-grade students whose teachers reported daily or weekly use of calculators had students with the highest scores, whereas fourth-grade teachers who reported that their students never used calculators had students with the lowest test scores. Other studies synthesizing and comparing research over the last two decades also support the fact that calculator and other computer technologies can enhance students' performance of arithmetical concepts and skills, as well as their problem-solving and higher-level abilities (Hembree and Dessart 1992; Burrill et al. 2002).

Q When should students master their basic computational facts?

A By the end of grade 2, students should know the basic addition and subtraction combinations, should be fluent in adding two-digit numbers, and should have methods for subtracting two-digit numbers. Students should know the basic multiplication and division combinations by the end of grade 4 and should be able to compute fluently with whole numbers by the end of grade 5.

Q How should students learn their basic single-digit facts?

A For most of the century, learning single-digit arithmetic has focused on memorization alone. However, basic-facts instruction that also emphasizes thinking strategies can make the tasks of learning new facts, recalling those facts, and learning additional material easier for students. As with any mathematical procedure, learning with understanding is more powerful than simply memorizing. Incorporating strategies built on number sense and meaningful mathematical relationships helps students develop well-understood meanings of the four basic operations. Students will also be able to carry these strategies over to multidigit computation problems. Some examples of these thinking strategies are as follows: 3×4 is the same as 4×3; 6×5 is 5 more than 5×5; 6×8 is double 3×8.

Q Should elementary schools use mathematics specialists?

A Some schools are using a model in which designated teachers assume responsibility for teaching mathematics to particular groups of students. Other schools identify mathematics teacher-leaders who work with teachers and organize professional development opportunities for the teachers in their school. Whatever model is used, a commitment to improving teachers' mathematical content and pedagogical knowledge in the elementary school is essential. Students in the elementary years need to be taught mathematics by teachers who have a deep understanding of the mathematics they teach and of how students learn mathematics, and who have efficient instructional and assessment strategies that help students learn mathematics.

Q What does algebra really entail in the elementary school?

A The word *algebra* is not commonly heard in elementary school classrooms, but the mathematical investigations and conversations of students in these grades frequently include elements of algebraic reasoning. Even before formal schooling, children develop beginning concepts related to patterns, functions, and other algebraic topics. They learn repetitive songs, rhythmic chants, and poems that are based on repeating and growing patterns. When students notice that operations seem to have particular properties, they are beginning to think algebraically. For example, students in prekindergarten through grade 2 may realize that changing the order in which two numbers are added does not change the result, or that adding zero to a number leaves that number unchanged. Students in grades 3–5 may describe in mathematical sentences the patterns they see in a display of "growing squares" (NCTM 2000, p. 159), shown below.

All these examples illustrate the building of algebraic understanding in the elementary school years.

Fig. 5.3.

Expressing "growing squares" in mathematical sentences (Adapted from Burton et al. 1992, p.6)

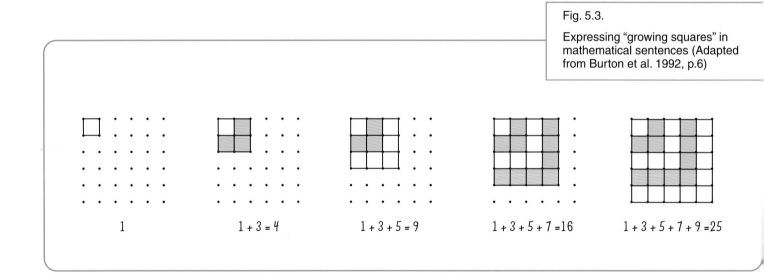

1   1 + 3 = 4   1 + 3 + 5 = 9   1 + 3 + 5 + 7 = 16   1 + 3 + 5 + 7 + 9 = 25

Q What about algebra in the eighth grade?

A All students should have a solid background in algebra by the end of eighth grade, whether or not it is obtained through a formal course. *Principles and Standards* (NCTM 2000) provides guidelines for introducing algebraic concepts early and then reinforcing and strengthening them throughout the grades. By viewing algebra as a strand in the curriculum from prekindergarten through grade 12, teachers can help students build a solid foundation of understanding and experience as a preparation for more-sophisticated work in algebra in the middle grades and high school. Specifically, the NCTM Standards for grades 6–8 focus on algebra, as well as its connection with other important areas, such as geometry.

Q Do all students need four years of high school mathematics?

A Yes. Whether students plan to pursue the further study of mathematics, to immediately enter the workforce, or to pursue other postsecondary education, they can be sure that mathematics will be a part of their future. A solid four years of secondary mathematics instruction in high school will give students an advantage no matter what track they pursue after high school.

Q How do I address those who may hold different views on teaching mathematics?

A As an administrator, you may be faced with parents, teachers, and others who have different views regarding teaching mathematics. Respond professionally, and be proactive. Help those who have opposing views see that you all share a concern for students and want them to be prepared for everyday life and for the workplace. Listen to the concerns of these critics. Provide facts and data to support the mathematics program and to counteract any misconceptions the critics may have. Engage in productive discussions, and find common ideas that you share with the critics.

Resources from NCTM

Principles and Standards Resources

Principles and Standards for School Mathematics (Principles and Standards)
A resource and guide for the mathematics education of students in prekindergarten through grade 12. Includes the E-Standards CD-ROM with the searchable full text and graphics of *Principles and Standards* and electronic examples for each grade band. Stock #719

Principles and Standards Overview
A booklet summarizing the main messages of *Principles and Standards*. Stock #737

Principles and Standards Outreach CD (2nd Edition)
Offers PowerPoint presentations, video clips, and additional handouts to support presentations and workshops on the messages of *Principles and Standards*. Stock #12260

Principles and Standards for School Mathematics: Vision to Reality
A forty-minute videotape that contains classroom clips and examines the major themes of *Principles and Standards* from each grade band. Stock #758

Frequently Asked Questions about "Principles and Standards for School Mathematics"
Excellent for distribution at meetings or for group discussions. Stock #12492; FREE

Illuminations (illuminations.nctm.org)
A Web site offering *Standards*-based lesson plans, including interactive applets for exploring, learning, and applying mathematics.

Navigations
A grade-band series of books in each of the major mathematics content areas, designed to give teachers and others suggestions, activities, and materials to put the ideas from *Principles and Standards* into classroom practice. A CD-ROM with additional resources, such as electronic applets and selected articles, accompanies each book.

Quick Reference Guide: Table of Standards and Expectations
Produced from the "Table of Standards and Expectations" in *Principles and Standards;* includes three 11-by-17-inch sheets to display the expectations across the four grade bands for each of the five content standards: Number and Operations, Algebra, Geometry, Data Analysis and Probability, and Measurement. Also summarizes the principles and process standards. Excellent resource for teachers and professional development workshops. Stock #12493

A Research Companion to "Principles and Standards for School Mathematics"
Contains research relevant to topics covered in *Principles and Standards* and supplies a comprehensive analysis of what research should be expected to do in setting standards for school mathematics. Stock #12341

Journals and Publications

NCTM publishes four professional journals: *Teaching Children Mathematics, Mathematics Teaching in the Middle School, Mathematics Teacher* and *Journal for Research in Mathematics Education.* NCTM also publishes the *NCTM News Bulletin,* a newsletter about events and issues affecting mathematics teachers, and more than 200 educational books, videotapes, and other materials.

Professional Development

The NCTM Academy for Professional Development
Offers grade-band Institutes designed for K–12 teachers of mathematics. Institute participants experience challenging and authentic mathematics investigations in-depth—as students—while presenters model current pedagogical techniques and new ways of thinking about teaching.

NCTM Annual Meeting and Regional Conferences
NCTM holds an annual meeting and several regional conferences each year, at which mathematics teachers and others interested in mathematics education can attend lectures, panel discussions, and workshops and can view exhibits of the latest mathematics education materials and innovations.

NCTM on the World Wide Web
For a complete list of resources and other information, visit the NCTM WEB site at www.nctm.org.

Other Resources

K–12 Mathematics Curriculum Center
National Science Foundation (NSF)-funded project that supports school districts as they build effective mathematics education programs that embrace the NCTM *Standards*. The Center provides a variety of resources, such as seminars, resource guides, and telephone consultations.
http://www.edc.org/mcc

Eisenhower National Clearinghouse for Mathematics and Science Education (ENC)
ENC's mission is to identify effective curriculum resources, create high-quality professional development materials, and disseminate useful information and products to improve K–12 mathematics and science teaching and learning.
http://www.enc.org

National Center for Improving Student Learning and Achievement in Mathematics and Science (NCISLA)
NCISLA is a university-based research center focusing on K–12 mathematics and science education. Through research and development, the Center is identifying new professional development models and ways that schools can better support teacher professional development and student learning.
http://www.wcer.wisc.edu/ncisla/index.html

Mathematical Sciences Education Board of the Center for Education (MSEB)
The MSEB is located within the National Research Council's Center for Education (CFE). Its mission is to provide national leadership and guidance for policies, programs, and practices supporting the improvement of mathematics education at all levels and for all members of society.
http://www.nationalacademies.org/mseb/index.html

AAAS Project 2061
Project 2061 is the long-term initiative of the American Association for the Advancement of Science working to reform K–12 science, mathematics, and technology education nationwide. Some of these initiatives include evaluating mathematics and science textbooks and assessments, creating conceptual strand maps, and leading workshops for educators.
http://www.project2061.org

The Math Forum
The Math Forum is a leading center for mathematics and mathematics education accessed on the Internet. The Math Forum's mission is to provide resources, materials, activities, person-to-person interactions, and educational products and services that enrich and support teaching and learning in an increasingly technological world. This online community includes teachers, students, researchers, parents, educators, and citizens at all levels who have an interest in mathematics and mathematics education.
http://mathforum.org

National Council of Supervisors of Mathematics (NCSM)
An affiliate of the NCTM, the NCSM supports mathematics education leadership at the school, district, college/university, state/province, and national levels.
http://www.ncsmonline.org

Table of NCTM Standards and Expectations

The Table of Standards and Expectations highlights the goals and expectations described in *Principles and Standards* for each of the four grade bands (pre-K–2, 3–5, 6–8, and 9–12) (NCTM 2000, pp. 392–402). The mathematics curriculum should be designed so that students build a successively deeper and more refined understanding of each of the Standards as they move through the grades.

The Content Standards—Number and Operations, Algebra, Geometry, Measurement, and Data Analysis and Probability—describe the mathematical content students should learn. The Process Standards—Problem Solving, Reasoning and Proof, Communication, Connections, and Representation—highlight ways of acquiring and using content knowledge. Together, these ten Standards define the basic mathematics that all students should have the opportunity to learn.

This table can be purchased separately as a resource for teachers and others. For more information, see Quick Reference Guide: Table of Standards and Expectations *on page 32 of this guide.*

Number and Operations

Source: *Principles and Standards for School Mathematics* (NCTM 2000, pp. 392–402)

STANDARD

Instructional programs from prekindergarten through grade 12 should enable all students to—

	Pre-K–2	Grades 3–5
	Expectations In prekindergarten through grade 2 all students should–	**Expectations** In grades 3–5 all students should–
Understand numbers, ways of representing numbers, relationships among numbers, and number systems	• count with understanding and recognize "how many" in sets of objects; • use multiple models to develop initial understandings of place value and the base-ten number system; • develop understanding of the relative position and magnitude of whole numbers and of ordinal and cardinal numbers and their connections; • develop a sense of whole numbers and represent and use them in flexible ways, including relating, composing, and decomposing numbers; • connect number words and numerals to the quantities they represent, using various physical models and representations; • understand and represent commonly used fractions, such as 1/4, 1/3, and 1/2.	• understand the place-value structure of the base-ten number system and be able to represent and compare whole numbers and decimals; • recognize equivalent representations for the same number and generate them by decomposing and composing numbers; • develop understanding of fractions as parts of unit wholes, as parts of a collection, as locations on number lines, and as divisions of whole numbers; • use models, benchmarks, and equivalent forms to judge the size of fractions; • recognize and generate equivalent forms of commonly used fractions, decimals, and percents; • explore numbers less than 0 by extending the number line and through familiar applications; • describe classes of numbers according to characteristics such as the nature of their factors.
Understand meanings of operations and how they relate to one another	• understand various meanings of addition and subtraction of whole numbers and the relationship between the two operations; • understand the effects of adding and subtracting whole numbers; • understand situations that entail multiplication and division, such as equal groupings of objects and sharing equally.	• understand various meanings of multiplication and division; • understand the effects of multiplying and dividing whole numbers; • identify and use relationships between operations, such as division as the inverse of multiplication, to solve problems; • understand and use properties of operations, such as the distributivity of multiplication over addition.
Compute fluently and make reasonable estimates	• develop and use strategies for whole-number computations, with a focus on addition and subtraction; • develop fluency with basic number combinations for addition and subtraction; • use a variety of methods and tools to compute, including objects, mental computation, estimation, paper and pencil, and calculators.	• develop fluency with basic number combinations for multiplication and division and use these combinations to mentally compute related problems, such as 30×50; • develop fluency in adding, subtracting, multiplying, and dividing whole numbers; • develop and use strategies to estimate the results of whole-number computations and to judge the reasonableness of such results; • develop and use strategies to estimate computations involving fractions and decimals in situations relevant to students' experience; • use visual models, benchmarks, and equivalent forms to add and subtract commonly used fractions and decimals; • select appropriate methods and tools for computing with whole numbers from among mental computation, estimation, calculators, and paper and pencil according to the context and nature of the computation and use the selected method or tool.

Number and Operations

STANDARD

Instructional programs from prekindergarten through grade 12 should enable all students to—

Grades 6–8

Expectations

In grades 6–8 all students should–

Grades 9–12

Expectations

In grades 9–12 all students should–

Understand numbers, ways of representing numbers, relationships among numbers, and number systems	• work flexibly with fractions, decimals, and percents to solve problems; • compare and order fractions, decimals, and percents efficiently and find their approximate locations on a number line; • develop meaning for percents greater than 100 and less than 1; • understand and use ratios and proportions to represent quantitative relationships; • develop an understanding of large numbers and recognize and appropriately use exponential, scientific, and calculator notation; • use factors, multiples, prime factorization, and relatively prime numbers to solve problems; • develop meaning for integers and represent and compare quantities with them.	• develop a deeper understanding of very large and very small numbers and of various representations of them; • compare and contrast the properties of numbers and number systems, including the rational and real numbers, and understand complex numbers as solutions to quadratic equations that do not have real solutions; • understand vectors and matrices as systems that have some of the properties of the real-number system; • use number-theory arguments to justify relationships involving whole numbers.
Understand meanings of operations and how they relate to one another	• understand the meaning and effects of arithmetic operations with fractions, decimals, and integers; • use the associative and commutative properties of addition and multiplication and the distributive property of multiplication over addition to simplify computations with integers, fractions, and decimals; • understand and use the inverse relationships of addition and subtraction, multiplication and division, and squaring and finding square roots to simplify computations and solve problems.	• judge the effects of such operations as multiplication, division, and computing powers and roots on the magnitudes of quantities; • develop an understanding of properties of, and representations for, the addition and multiplication of vectors and matrices; • develop an understanding of permutations and combinations as counting techniques.
Compute fluently and make reasonable estimates	• select appropriate methods and tools for computing with fractions and decimals from among mental computation, estimation, calculators or computers, and paper and pencil, depending on the situation, and apply the selected methods; • develop and analyze algorithms for computing with fractions, decimals, and integers and develop fluency in their use; • develop and use strategies to estimate the results of rational-number computations and judge the reasonableness of the results; • develop, analyze, and explain methods for solving problems involving proportions, such as scaling and finding equivalent ratios.	• develop fluency in operations with real numbers, vectors, and matrices, using mental computation or paper-and-pencil calculations for simple cases and technology for more-complicated cases. • judge the reasonableness of numerical computations and their results.

Algebra
Standard

Instructional programs from prekindergarten through grade 12 should enable all students to—

	Pre-K–2 **Expectations** In prekindergarten through grade 2 all students should—	Grades 3–5 **Expectations** In grades 3–5 all students should—
Understand patterns, relations, and functions	• sort, classify, and order objects by size, number, and other properties; • recognize, describe, and extend patterns such as sequences of sounds and shapes or simple numeric patterns and translate from one representation to another; • analyze how both repeating and growing patterns are generated.	• describe, extend, and make generalizations about geometric and numeric patterns; • represent and analyze patterns and functions, using words, tables, and graphs.
Represent and analyze mathematical situations and structures using algebraic symbols	• illustrate general principles and properties of operations, such as commutativity, using specific numbers; • use concrete, pictorial, and verbal representations to develop an understanding of invented and conventional symbolic notations.	• identify such properties as commutativity, associativity, and distributivity and use them to compute with whole numbers; • represent the idea of a variable as an unknown quantity using a letter or a symbol; • express mathematical relationships using equations.
Use mathematical models to represent and understand quantitative relationships	• model situations that involve the addition and subtraction of whole numbers, using objects, pictures, and symbols.	• model problem situations with objects and use representations such as graphs, tables, and equations to draw conclusions.
Analyze change in various contexts	• describe qualitative change, such as a student's growing taller; • describe quantitative change, such as a student's growing two inches in one year.	• investigate how a change in one variable relates to a change in a second variable; • identify and describe situations with constant or varying rates of change and compare them.

Algebra
STANDARD

Instructional programs from prekindergarten through grade 12 should enable all students to—

Grades 6–8
Expectations
In grades 6–8 all students should–

Grades 9–12
Expectations
In grades 9–12 all students should–

Understand patterns, relations, and functions	• represent, analyze, and generalize a variety of patterns with tables, graphs, words, and, when possible, symbolic rules; • relate and compare different forms of representation for a relationship; • identify functions as linear or nonlinear and contrast their properties from tables, graphs, or equations.	• generalize patterns using explicitly defined and recursively defined functions; • understand relations and functions and select, convert flexibly among, and use various representations for them; • analyze functions of one variable by investigating rates of change, intercepts, zeros, asymptotes, and local and global behavior; • understand and perform transformations such as arithmetically combining, composing, and inverting commonly used functions, using technology to perform such operations on more-complicated symbolic expressions; • understand and compare the properties of classes of functions, including exponential, polynomial, rational, logarithmic, and periodic functions; • interpret representations of functions of two variables.
Represent and analyze mathematical situations and structures using algebraic symbols	• develop an initial conceptual understanding of different uses of variables; • explore relationships between symbolic expressions and graphs of lines, paying particular attention to the meaning of intercept and slope; • use symbolic algebra to represent situations and to solve problems, especially those that involve linear relationships; • recognize and generate equivalent forms for simple algebraic expressions and solve linear equations.	• understand the meaning of equivalent forms of expressions, equations, inequalities, and relations; • write equivalent forms of equations, inequalities, and systems of equations and solve them with fluency—mentally or with paper and pencil in simple cases and using technology in all cases; • use symbolic algebra to represent and explain mathematical relationships; • use a variety of symbolic representations, including recursive and parametric equations, for functions and relations; • judge the meaning, utility, and reasonableness of the results of symbol manipulations, including those carried out by technology.
Use mathematical models to represent and understand quantitative relationships	• model and solve contextualized problems using various representations, such as graphs, tables, and equations.	• identify essential quantitative relationships in a situation and determine the class or classes of functions that might model the relationships; • use symbolic expressions, including iterative and recursive forms, to represent relationships arising from various contexts; • draw reasonable conclusions about a situation being modeled.
Analyze change in various contexts	• use graphs to analyze the nature of changes in quantities in linear relationships.	• approximate and interpret rates of change from graphical and numerical data.

Geometry

STANDARD

Instructional programs from prekindergarten through grade 12 should enable all students to—

	Pre-K–2 Expectations In prekindergarten through grade 2 all students should—	Grades 3–5 Expectations In grades 3–5 all students should—
Analyze characteristics and properties of two- and three-dimensional geometric shapes and develop mathematical arguments about geometric relationships	• recognize, name, build, draw, compare, and sort two- and three-dimensional shapes; • describe attributes and parts of two- and three-dimensional shapes; • investigate and predict the results of putting together and taking apart two- and three-dimensional shapes.	• identify, compare, and analyze attributes of two- and three-dimensional shapes and develop vocabulary to describe the attributes; • classify two- and three-dimensional shapes according to their properties and develop definitions of classes of shapes such as triangles and pyramids; • investigate, describe, and reason about the results of subdividing, combining, and transforming shapes; • explore congruence and similarity; • make and test conjectures about geometric properties and relationships and develop logical arguments to justify conclusions.
Specify locations and describe spatial relationships using coordinate geometry and other representational systems	• describe, name, and interpret relative positions in space and apply ideas about relative position; • describe, name, and interpret direction and distance in navigating space and apply ideas about direction and distance; • find and name locations with simple relationships such as "near to" and in coordinate systems such as maps.	• describe location and movement using common language and geometric vocabulary; • make and use coordinate systems to specify locations and to describe paths; • find the distance between points along horizontal and vertical lines of a coordinate system.
Apply transformations and use symmetry to analyze mathematical situations	• recognize and apply slides, flips, and turns; • recognize and create shapes that have symmetry.	• predict and describe the results of sliding, flipping, and turning two-dimensional shapes; • describe a motion or a series of motions that will show that two shapes are congruent; • identify and describe line and rotational symmetry in two- and three-dimensional shapes and designs.
Use visualization, spatial reasoning, and geometric modeling to solve problems	• create mental images of geometric shapes using spatial memory and spatial visualization; • recognize and represent shapes from different perspectives; • relate ideas in geometry to ideas in number and measurement; • recognize geometric shapes and structures in the environment and specify their location	• build and draw geometric objects; • create and describe mental images of objects, patterns, and paths; • identify and build a three-dimensional object from two-dimensional representations of that object; • identify and draw a two-dimensional representation of a three-dimensional object; • use geometric models to solve problems in other areas of mathematics, such as number and measurement; • recognize geometric ideas and relationships and apply them to other disciplines and to problems that arise in the classroom or in everyday life.

Geometry
STANDARD

Instructional programs from prekindergarten through grade 12 should enable all students to—

Grades 6–8
Expectations

In grades 6–8 all students should–

Grades 9–12
Expectations

In grades 9–12 all students should–

Analyze characteristics and properties of two- and three-dimensional geometric shapes and develop mathematical arguments about geometric relationships	• precisely describe, classify, and understand relationships among types of two- and three-dimensional objects using their defining properties; • understand relationships among the angles, side lengths, perimeters, areas, and volumes of similar objects; • create and critique inductive and deductive arguments concerning geometric ideas and relationships, such as congruence, similarity, and the Pythagorean relationship.	• analyze properties and determine attributes of two- and three-dimensional objects; • explore relationships (including congruence and similarity) among classes of two- and three-dimensional geometric objects, make and test conjectures about them, and solve problems involving them; • establish the validity of geometric conjectures using deduction, prove theorems, and critique arguments made by others; • use trigonometric relationships to determine lengths and angle measures.
Specify locations and describe spatial relationships using coordinate geometry and other representational systems	• use coordinate geometry to represent and examine the properties of geometric shapes; • use coordinate geometry to examine special geometric shapes, such as regular polygons or those with pairs of parallel or perpendicular sides.	• use Cartesian coordinates and other coordinate systems, such as navigational, polar, or spherical systems, to analyze geometric situations; • investigate conjectures and solve problems involving two- and three-dimensional objects represented with Cartesian coordinates.
Apply transformations and use symmetry to analyze mathematical situations	• describe sizes, positions, and orientations of shapes under informal transformations such as flips, turns, slides, and scaling; • examine the congruence, similarity, and line or rotational symmetry of objects using transformations.	• understand and represent translations, reflections, rotations, and dilations of objects in the plane by using sketches, coordinates, vectors, function notation, and matrices; • use various representations to help understand the effects of simple transformations and their compositions.
Use visualization, spatial reasoning, and geometric modeling to solve problems	• draw geometric objects with specified properties, such as side lengths or angle measures; • use two-dimensional representations of three-dimensional objects to visualize and solve problems such as those involving surface area and volume; • use visual tools such as networks to represent and solve problems; • use geometric models to represent and explain numerical and algebraic relationships; • recognize and apply geometric ideas and relationships in areas outside the mathematics classroom, such as art, science, and everyday life.	• draw and construct representations of two- and three-dimensional geometric objects using a variety of tools; • visualize three-dimensional objects from different perspectives and analyze their cross sections; • use vertex-edge graphs to model and solve problems; • use geometric models to gain insights into, and answer questions in, other areas of mathematics; • use geometric ideas to solve problems in, and gain insights into, other disciplines and other areas of interest such as art and architecture.

Measurement

STANDARD

Instructional programs from prekindergarten through grade 12 should enable all students to—

	Pre-K–2	Grades 3–5
	Expectations	**Expectations**
	In prekindergarten through grade 2 all students should—	In grades 3–5 all students should—
Understand measurable attributes of objects and the units, systems, and processes of measurement	• recognize the attributes of length, volume, weight, area, and time; • compare and order objects according to these attributes; • understand how to measure using nonstandard and standard units; • select an appropriate unit and tool for the attribute being measured.	• understand such attributes as length, area, weight, volume, and size of angle and select the appropriate type of unit for measuring each attribute; • understand the need for measuring with standard units and become familiar with standard units in the customary and metric systems; • carry out simple unit conversions, such as from centimeters to meters, within a system of measurement; • understand that measurements are approximations and understand how differences in units affect precision; • explore what happens to measurements of a two-dimensional shape such as its perimeter and area when the shape is changed in some way.
Apply appropriate techniques, tools, and formulas to determine measurements	• measure with multiple copies of units of the same size, such as paper clips laid end to end; • use repetition of a single unit to measure something larger than the unit, for instance, measuring the length of a room with a single meterstick; • use tools to measure; • develop common referents for measures to make comparisons and estimates.	• develop strategies for estimating the perimeters, areas, and volumes of irregular shapes; • select and apply appropriate standard units and tools to measure length, area, volume, weight, time, temperature, and the size of angles; • select and use benchmarks to estimate measurements; • develop, understand, and use formulas to find the area of rectangles and related triangles and parallelograms; • develop strategies to determine the surface areas and volumes of rectangular solids.

Measurement

Standard

Instructional programs from prekindergarten through grade 12 should enable all students to—

	Grades 6–8	Grades 9–12
	Expectations	**Expectations**
	In grades 6–8 all students should–	In grades 9–12 all students should–
Understand measurable attributes of objects and the units, systems, and processes of measurement	• understand both metric and customary systems of measurement; • understand relationships among units and convert from one unit to another within the same system; • understand, select, and use units of appropriate size and type to measure angles, perimeter, area, surface area, and volume.	• make decisions about units and scales that are appropriate for problem situations involving measurement.
Apply appropriate techniques, tools, and formulas to determine measurements	• use common benchmarks to select appropriate methods for estimating measurements; • select and apply techniques and tools to accurately find length, area, volume, and angle measures to appropriate levels of precision; • develop and use formulas to determine the circumference of circles and the area of triangles, parallelograms, trapezoids, and circles and develop strategies to find the area of more-complex shapes; • develop strategies to determine the surface area and volume of selected prisms, pyramids, and cylinders; • solve problems involving scale factors, using ratio and proportion; • solve simple problems involving rates and derived measurements for such attributes as velocity and density.	• analyze precision, accuracy, and approximate error in measurement situations; • understand and use formulas for the area, surface area, and volume of geometric figures, including cones, spheres, and cylinders; • apply informal concepts of successive approximation, upper and lower bounds, and limit in measurement situations; • use unit analysis to check measurement computations.

Data Analysis and Probability

STANDARD

Instructional programs from prekindergarten through grade 12 should enable all students to—

	Pre-K–2	Grades 3–5
	Expectations In prekindergarten through grade 2 all students should–	**Expectations** In grades 3–5 all students should–
Formulate questions that can be addressed with data and collect, organize, and display relevant data to answer them	• pose questions and gather data about themselves and their surroundings; • sort and classify objects according to their attributes and organize data about the objects; • represent data using concrete objects, pictures, and graphs.	• design investigations to address a question and consider how data-collection methods affect the nature of the data set; • collect data using observations, surveys, and experiments; • represent data using tables and graphs such as line plots, bar graphs, and line graphs; • recognize the differences in representing categorical and numerical data.
Select and use appropriate statistical methods to analyze data	• describe parts of the data and the set of data as a whole to determine what the data show.	• describe the shape and important features of a set of data and compare related data sets, with an emphasis on how the data are distributed; • use measures of center, focusing on the median, and understand what each does and does not indicate about the data set; • compare different representations of the same data and evaluate how well each representation shows important aspects of the data.
Develop and evaluate inferences and predictions that are based on data	• discuss events related to students' experiences as likely or unlikely.	• propose and justify conclusions and predictions that are based on data and design studies to further investigate the conclusions or predictions.
Understand and apply basic concepts of probability		• describe events as likely or unlikely and discuss the degree of likelihood using such words as *certain*, *equally likely*, and *impossible*; • predict the probability of outcomes of simple experiments and test the predictions; • understand that the measure of the likelihood of an event can be represented by a number from 0 to 1.

Data Analysis and Probability

STANDARD

Instructional programs from prekindergarten through grade 12 should enable all students to—

	Grades 6–8 Expectations In grades 6–8 all students should–	Grades 9–12 Expectations In grades 9–12 all students should–
Formulate questions that can be addressed with data and collect, organize, and display relevant data to answer them	• formulate questions, design studies, and collect data about a characteristic shared by two populations or different characteristics within one population; • select, create, and use appropriate graphical representations of data, including histograms, box plots, and scatterplots.	• understand the differences among various kinds of studies and which types of inferences can legitimately be drawn from each; • know the characteristics of well-designed studies, including the role of randomization in surveys and experiments; • understand the meaning of measurement data and categorical data, of univariate and bivariate data, and of the term *variable;* • understand histograms, parallel box plots, and scatterplots and use them to display data; • compute basic statistics and understand the distinction between a statistic and a parameter.
Select and use appropriate statistical methods to analyze data	• find, use, and interpret measures of center and spread, including mean and interquartile range; • discuss and understand the correspondence between data sets and their graphical representations, especially histograms, stem-and-leaf plots, box plots, and scatterplots.	• for univariate measurement data, be able to display the distribution, describe its shape, and select and calculate summary statistics; • for bivariate measurement data, be able to display a scatterplot, describe its shape, and determine regression coefficients, regression equations, and correlation coefficients using technological tools; • display and discuss bivariate data where at least one variable is categorical; • recognize how linear transformations of univariate data affect shape, center, and spread; • identify trends in bivariate data and find functions that model the data or transform the data so that they can be modeled.
Develop and evaluate inferences and predictions that are based on data	• use observations about differences between two or more samples to make conjectures about the populations from which the samples were taken; • make conjectures about possible relationships between two characteristics of a sample on the basis of scatterplots of the data and approximate lines of fit; • use conjectures to formulate new questions and plan new studies to answer them.	• use simulations to explore the variability of sample statistics from a known population and to construct sampling distributions; • understand how sample statistics reflect the values of population parameters and use sampling distributions as the basis for informal inference; • evaluate published reports that are based on data by examining the design of the study, the appropriateness of the data analysis, and the validity of conclusions; • understand how basic statistical techniques are used to monitor process characteristics in the workplace.
Understand and apply basic concepts of probability	• understand and use appropriate terminology to describe complementary and mutually exclusive events; • use proportionality and a basic understanding of probability to make and test conjectures about the results of experiments and simulations; • compute probabilities for simple compound events, using such methods as organized lists, tree diagrams, and area models.	• understand the concepts of sample space and probability distribution and construct sample spaces and distributions in simple cases; • use simulations to construct empirical probability distributions; • compute and interpret the expected value of random variables in simple cases; • understand the concepts of conditional probability and independent events; • understand how to compute the probability of a compound event.

Problem Solving

STANDARD

Instructional programs from prekindergarten through grade 12 should enable all students to—

- Build new mathematical knowledge through problem solving
- Solve problems that arise in mathematics and in other contexts
- Apply and adapt a variety of appropriate strategies to solve problems
- Monitor and reflect on the process of mathematical problem solving

Reasoning and Proof

STANDARD

Instructional programs from prekindergarten through grade 12 should enable all students to—

- Recognize reasoning and proof as fundamental aspects of mathematics
- Make and investigate mathematical conjectures
- Develop and evaluate mathematical arguments and proofs
- Select and use various types of reasoning and methods of proof

Communication

STANDARD

Instructional programs from prekindergarten through grade 12 should enable all students to—

- Organize and consolidate their mathematical thinking through communication
- Communicate their mathematical thinking coherently and clearly to peers, teachers, and others
- Analyze and evaluate the mathematical thinking and strategies of others
- Use the language of mathematics to express mathematical ideas precisely

Connections

STANDARD

Instructional programs from prekindergarten through grade 12 should enable all students to—

- Recognize and use connections among mathematical ideas
- Understand how mathematical ideas interconnect and build on one another to produce a coherent whole
- Recognize and apply mathematics in contexts outside of mathematics

Representation

STANDARD

Instructional programs from prekindergarten through grade 12 should enable all students to—

- Create and use representations to organize, record, and communicate mathematical ideas
- Select, apply, and translate among mathematical representations to solve problems
- Use representations to model and interpret physical, social, and mathematical phenomena

REFERENCES

Bennett, Albert, Eugene Maier, and Ted Nelson. *Math and the Mind's Eye.* Portland, Oreg.: The Math Learning Center, 1988–1998.

Black, Paul, and Dylan Wiliam. "Inside the Black Box: Raising Standards through Classroom Assessment." *Phi Delta Kappan* (October 1998): 139–48.

Boers-van Oosterum, Monique Agnes Maria. "Understanding of Variables and Their Uses Acquired by Students in Traditional and Computer-Intensive Algebra." Ph.D. diss., University of Maryland—College Park, 1990.

Bransford, John D., Ann L. Brown, and Rodney R. Cocking, eds. *How People Learn: Brain, Mind, Experience, and School.* Washington, D.C.: National Academy Press, 1999.

Brown, Catherine A., and Margaret S. Smith. "Supporting the Development of Mathematical Pedagogy." *Mathematics Teacher* 90 (February 1997): 138–43.

Brownell, William A. "The Place of Meaning in the Teaching of Arithmetic." *Elementary School Journal* 47 (January 1947): 256–65.

Burrill, Gail, et al. *Handheld Graphing Technology in Secondary Mathematics: Research Findings and Implications for Classroom Practice.* Report prepared through a grant to Michigan State University. Dallas, Tex.: Texas Instruments, 2002.

Campbell, Patricia. *Project IMPACT: Increasing Mathematics Power for All Children and Teachers.* Phase 1, Final Report. College Park, Md.: Center for Mathematics Education, University of Maryland, 1995.

Dunham, Penelope H., and Thomas P. Dick. "Research on Graphing Calculators." *Mathematics Teacher* 87 (September 1994): 440–45.

Fey, James. "Mathematics Teaching Today: Perspectives from Three National Surveys." *Mathematics Teacher* 72 (1979): 490–504.

Fuson, Karen C., James W. Stigler, and Karen Bartsch. "Brief Report: Grade Placement of Addition and Subtraction Topics in Japan, Mainland China, the Soviet Union, Taiwan, and the United States." *Journal for Research in Mathematics Education* 19 (1988): 449–56.

Gamoran, Adam, Charles W. Anderson, Pamela Anne Quiroz, Walter G. Secada, Tona Williams, and Scott Ashmann. *Transforming Teaching in Math and Science: How Schools and Districts Can Support Change.* New York: Teachers College Press, in press.

Griffin, Sharon A., Robbie Case, and Robert S. Siegler. "Right-Start: Providing the Central Conceptual Prerequisites for First Formal Learning of Arithmetic to Students at Risk for School Failure." In *Classroom Lessons: Integrating Cognitive Theory and Classroom Practice,* edited by Kate McGilly, pp. 25–49. Cambridge: MIT Press, 1994.

Groves, Susie. "Calculators: A Learning Environment to Promote Number Sense." Paper presented at the annual meeting of the American Educational Research Association, New Orleans, April 1994.

Hecker, Daniel E. "Occupational Employment Projections to 2010." In *Monthly Labor Review* 124 (11) (November 2001): 57–84.

Hembree, Ray, and Donald J. Dessart. "Research on Calculators in Mathematics Education." In *Calculators in Mathematics Education,* 1992 Yearbook of the National Council of Teachers of Mathematics (NCTM), edited by James T. Fey, pp. 23–32. Reston, Va.: NCTM, 1992.

Hiebert, James, ed. *Conceptual and Procedural Knowledge: The Case of Mathematics.* Hillsdale, N.J.: Lawrence Erlbaum Associates, 1986.

Hiebert, James, and Thomas P. Carpenter. "Learning and Teaching with Understanding." In *Handbook of Research on Mathematics Teaching and Learning*, edited by Douglas A. Grouws, pp.65–97. New York: Macmillan Publishing Co., 1992.

Hoetker, J., and W. Ahlbrand. "The Persistence of the Recitation." *American Educational Research Journal* 6 (1969):145–67

Knapp, Michael S., Nancy E. Adelman, Camille Marder, Heather McCollum, Margaret C. Needels, Christine Padilla, Patrick M. Shields, Brenda J. Turnbull, and Andrew A. Zucker. *Teaching for Meaning in High-Poverty Schools.* New York: Teachers College Press, 1995.

Kouba, Vicky L., and Diana Wearne. "Whole Number Properties and Operations." In *Results from the Seventh Mathematics Assessment of the National Assessment of Educational Progress*, edited by Edward A. Silver and Patricia A. Kenney, pp. 141–61. Reston, Va.: National Council of Teachers of Mathematics, 2000.

McKnight, Curtis C., F. Joe Crosswhite, John A. Dossey, Edward Kifer, Jane O. Swafford, Kenneth T. Travers, and Thomas J. Cooney. *The Underachieving Curriculum: Assessing U.S. School Mathematics from an International Perspective.* Champaign, Ill.: Stipes Publishing Co., 1987.

McKnight, Curtis C., and William H. Schmidt. "Facing Facts in U.S. Science and Mathematics Education: Where We Stand, Where We Want to Go. *Journal of Science Education and Technology* 7 (1): 57–76, 1998.

Ma, Liping. *Knowing and Teaching Elementary Mathematics: Teachers' Understanding of Fundamental Mathematics in China and the United States.* Mahwah, N.J.: Lawrence Erlbaum Associates, 1999.

National Advisory Committee on Mathematics Education. *Overview and Analysis of School Mathematics, Grades K–12.* Washington, D.C.: Conference Board of the Mathematical Sciences, 1975.

National Center for Education Statistics. "National Assessment of Educational Progress (NAEP), 1990, 1992, 1996, and 2000 Mathematics Assessments." http://nces.ed.gov/nationsreportcard/mathematics/results [cited May 1, 2003a]. World Wide Web.

———. http://nces.ed.gov/timss/highlights.asp. "Trends in International Mathematics and Science Study TIMSS)—Case Study" [cited May 1, 2003b]. World Wide Web.

———. http://nces.ed.gov/timss/fig1.asp. "Trends in International Mathematics and Science Study TIMSS)—Video Studies" [cited May 1, 2003c]. World Wide Web.

National Commission on Mathematics and Science Teaching for the Twenty-first Century. *Before It's Too Late*: *A Report from the National Commission on Mathematics and Science Teaching for the Twenty-first Century.* Washington, D.C.: U.S. Department of Education, September 2003.

National Council of Teachers of Mathematics (NCTM). *Professional Standards for Teaching Mathematics.* Reston, Va.: NCTM, 1991.

———. *Assessment Standards for School Mathematics.* Reston, Va.: NCTM, 1995.

———. *Principles and Standards for School Mathematics.* Reston, Va.: NCTM, 2000.

National Research Council. *Adding It Up: Helping Children Learn Mathematics,* edited by Jeremy Kilpatrick, Jane Swafford, and Bradford Findell. Mathematics Learning Study Committee, Center for Education, Division of Behavioral and Social Sciences and Education. Washington, D.C.: National Academy Press, 2001.

Peak, Lois. *Pursuing Excellence: A Study of U.S. Eighth-grade Mathematics and Science Teaching, Learning, Curriculum, and Achievement in an International Context.* Washington, D.C.: National Center for Educational Statistics, 1996.

Putnam, Ralph T., and Hilda Borko. "What Do New Views of Knowledge and Thinking Have to Say about Research on Teacher Learning?" *Educational Researcher* 29 (January–February 2000): 4–15.

Rojano, Teresa. "Developing Algebraic Aspects of Problem Solving within a Spreadsheet Environment." In *Approaches to Algebra: Perspectives for Research and Teaching,* edited by Nadine Bednarz, Carolyn Kieran, and Lesley Lee. Boston: Kluwer Academic Publishers, 1996.

Silver, Edward A., and Mary Kay Stein. "The QUASAR Project: The 'Revolution of the Possible' in Mathematics Instructional Reform in Urban Middle Schools." *Urban Education* 30 (January 1996): 476–521.

Skemp, Richard R. "Relational Understanding and Instrumental Understanding." *Mathematics Teaching* 77 (December 1976): 20–26. Reprinted in *Arithmetic Teacher* 26 (November 1978): 9–15.

Smith, Margaret S. "Balancing Old and New: An Experienced Middle School Teacher's Learning in the Context of Mathematics Instructional Reform." *Elementary School Journal* 100 (4) (2000): 351–76.

Stigler, James W., and James Hiebert. "Understanding and Improving Classroom Mathematics Instruction: An Overview of the TIMSS Video Study." *Phi Delta Kappan* 79 (1) (1997): 14–21.

———. *The Teaching Gap: Best Ideas from the World's Teachers for Improving Education in the Classroom.* New York: The Free Press, 1999.

Tharp, Roland G., and Ronald Gallimore. *Rousing Minds to Life: Teaching, Learning, and Schooling in Social Context.* New York: Cambridge University Press, 1988.

Wearne, Diana, and Vicky L. Kouba. "Rational Numbers." In *Results from the Seventh Mathematics Assessment of the National Assessment of Educational Progress,* edited by Edward A. Silver and Patricia A. Kenney pp. 163–91. Reston, Va.: National Council of Teachers of Mathematics, 2000.

About the National Council of Teachers of Mathematics

NCTM is a public voice of mathematics education, providing vision, leadership, and professional development to support teachers in ensuring mathematics learning of the highest quality for all students.

In addition to *Principles and Standards for School Mathematics*, published in 2000, NCTM publishes the Navigations series, which translates the Principles and Standards into action by highlighting major mathematics content areas and making them classroom friendly in grade-band specific volumes.

Visit www.nctm.org for information on NCTM membership, as well as NCTM's journals and books, online resources, and professional development conferences and workshops.